Atlas of Plant Cell Structure

Tetsuko Noguchi
Editor-in-Chief

Shigeyuki Kawano • Hirokazu Tsukaya
Sachihiro Matsunaga • Atsushi Sakai
Ichirou Karahara • Yasuko Hayashi
Editors

Atlas of Plant Cell Structure

Editor-in-Chief
Tetsuko Noguchi
Professor, Course of Biological Sciences
Faculty of Science
Nara Women's University
Kitauoya-nishimachi, Nara, Japan

Editors

Shigeyuki Kawano
Professor, Department of Integrated
  Biosciences
Graduate School of Frontier Sciences
The University of Tokyo
Kashiwa, Chiba, Japan

Sachihiro Matsunaga
Professor, Department of Applied
  Biological Science
Faculty of Science and Technology
Tokyo University of Science
Noda, Chiba, Japan

Ichirou Karahara
Professor, Department of Biology
Graduate School of Science
  and Engineering
University of Toyama
Toyama, Toyama, Japan

Hirokazu Tsukaya
Professor, Department of Biological Sciences
Graduate School of Science
The University of Tokyo
Bunkyo-ku, Tokyo, Japan

Atsushi Sakai
Professor, Course of Biological Sciences
Faculty of Science
Nara Women's University
Kitauoya-nishimachi, Nara, Japan

Yasuko Hayashi
Associate Professor, Department of
  Environmental Science and Technology
Graduate School of Science and Technology
Niigata University
Ikarashi, Niigata, Japan

ISBN 978-4-431-54940-6   ISBN 978-4-431-54941-3 (eBook)
DOI 10.1007/978-4-431-54941-3
Springer Tokyo Heidelberg New York Dordrecht London

Library of Congress Control Number: 2014942687

© Springer Japan 2014
This work is subject to copyright. All rights are reserved by the Publisher, whether the whole or part of the material is concerned, specifically the rights of translation, reprinting, reuse of illustrations, recitation, broadcasting, reproduction on microfilms or in any other physical way, and transmission or information storage and retrieval, electronic adaptation, computer software, or by similar or dissimilar methodology now known or hereafter developed. Exempted from this legal reservation are brief excerpts in connection with reviews or scholarly analysis or material supplied specifically for the purpose of being entered and executed on a computer system, for exclusive use by the purchaser of the work. Duplication of this publication or parts thereof is permitted only under the provisions of the Copyright Law of the Publisher's location, in its current version, and permission for use must always be obtained from Springer. Permissions for use may be obtained through RightsLink at the Copyright Clearance Center. Violations are liable to prosecution under the respective Copyright Law.
The use of general descriptive names, registered names, trademarks, service marks, etc. in this publication does not imply, even in the absence of a specific statement, that such names are exempt from the relevant protective laws and regulations and therefore free for general use.
While the advice and information in this book are believed to be true and accurate at the date of publication, neither the authors nor the editors nor the publisher can accept any legal responsibility for any errors or omissions that may be made. The publisher makes no warranty, express or implied, with respect to the material contained herein.

Printed on acid-free paper

Springer is part of Springer Science+Business Media (www.springer.com)

Preface

This is a photo catalog of the world's cellular diversity, which plant morphologists have been studying for over a quarter of a century. We assembled this atlas for science students, their teachers, and anyone who is curious about the extraordinary variety of living things studied in the field of plant morphology. Much of the knowledge described here, particularly about flowering plants, mosses, liverworts, algae, fungi, and lichens, has been gathered only in the past quarter century and represents the frontiers of research.

"Seeing is believing" is an idiom first recorded in 1639 that means "only physical or concrete evidence is convincing" and is a popular saying throughout the world. It is extremely difficult to describe in full any given state via text alone. Properly shot photos, on the other hand, can show such states with vastly more precision and detail. The saying leads to the mistaken belief that "seen evidence" can be easily and correctly interpreted, when in fact, interpretation may be difficult. An advanced transmission electron microscopy (TEM) technology, 3D images reconstructed from a series of TEM images, provides many insights into structural differences of organelles in a single cell and between cell types. As further evidence that seeing is believing, fluorescence microscopy is an excellent methodology for analyzing mitochondrial dynamics, for instance. A particularly important methodology is to use fluorescent dyes to stain special cell structures in many types of living cells, either by natural affinity or by tagging. Today, the use of green fluorescent protein (GFP) for visualizing a particular protein in living cells has been applicable to the research of cell organelle dynamics. Some variants of GFP contain unique optical properties which can provide marvelous, elegant experiments and lead to breakthrough findings by cell biologists.

Morphological research on plant bodies and their structures began in ancient Greece over 2,000 years ago. It is the biologist's drive to see and learn more about the formation of individual plants and organs that has paved the way to various microscopes and visualization techniques. In the seventeenth century, the light microscope was invented, which led to Robert Hooke's presenting his observations of what he called "cells" in cork, a dead plant organ, in 1655. In the nineteenth century, nuclei, mitochondria, chloroplasts, and vacuoles were identified in living cells, and 46 years after the first observation of the nucleus, the behavior of chromosomes during mitosis was reported. With the development of electron microscopy, biologists came to understand that individual organelles have the same basic ultrastructure, and discovered that organelles such as the endoplasmic reticulum, plant Golgi body, and cytoskeleton microtubules and actin filaments are common to all vegetative cells. Today, the resolution of these microscopes is greatly advanced and the techniques for specimen preparation have improved. With transgenic techniques, high-resolution fluorescent microscopy, and so on, we can detect not only the behaviors of organelles but also the dynamic movements of molecules to carry out a variety of functions that are critical for life.

The main concept of this book is visualization: seeing is believing. It presents beautiful photographs and 3D-reconstruction images of cellular structures in plants, algae, fungi, and related organisms taken by a variety of microscopes and visualization techniques. The book is subdivided into nine chapters: 1. Nuclei and Chromosomes, 2. Mitochondria, 3. Chloroplasts, 4. The Endoplasmic Reticulum, Golgi Apparatuses, and Endocytic Organelles, 5. Vacuoles and Storage Organelles, 6. Cytoskeletons, 7. Cell Walls, 8. Generative Cells, and 9. Meristems.

The topics in each chapter are restricted with the hope that they will be interesting, useful, and comprehensible. Each photo plate is accompanied by an explanation to introduce readers to the cell morphology, structures, and functions depicted. The explanations are intended to be the minimum necessary, because we hope readers will deeply consider and understand the minute ultrastructure within cells directly from the photographs. The references that are included will help readers to understand the topics in depth, but references are avoided in some photo plates showing popular cell structures.

This *Atlas of Plant Cell Structure* is published to celebrate the 25th anniversary of the Japanese Society of Plant Morphology. The Society was founded in 1988 to promote studies of plant morphology and discussion among researchers in this academic field. The members are scientists in plant morphology (micro- and macro-structure, function, and development), all levels of organization (molecular to macro-structure), and all plant groups and related organisms (cyanobacteria, algae, fungi, and lichens).

Included are 92 beautiful photographs, contributed mostly by members of the Society. Some of the photographs are originals, made especially for this atlas, and some have been featured on the cover of journals such as *Plant Morphology* (the official journal of the Society), the *Journal of Cell Science*, the *Journal of Plant Research*, *Molecular Biology of the Cell*, and *Nature*. We hope readers will enjoy this visual tour within cells and get new insights into plant cell structure.

Nara, Japan	Tetsuko Noguchi
Chiba, Japan	Shigeyuki Kawano
Tokyo, Japan	Hirokazu Tsukaya
Chiba, Japan	Sachihiro Matsunaga
Nara, Japan	Atsushi Sakai
Toyama, Japan	Ichirou Karahara
Niigata, Japan	Yasuko Hayashi

September 2014

Acknowledgments

The photographs in this book were taken by 66 members of the Japanese Society of Plant Morphology using the best methods and techniques available to capture the objects in cells and tissues: by light and fluorescent microscopy, by transmission and scanning electron microscopy, and by other processes. In addition, 25 non-members of the society kindly agreed to allow their photographs to be reproduced in the book.

Morphologists recognize how important it is for them to observe objects closely. In addition, they always try to capture real images of cell structures that reflect active functions being carried out by complex molecular mechanisms. The photographs in this book were taken by such plant morphologists.

We deeply thank Reiko Suzuki and Kayoko Morita of Nara Women's University for their help in digital processing of the images and preparation of the manuscripts for this book.

We extend our thanks to all those who have encouraged the publication of this atlas.

September 2014

Contents

1 **Nuclei and Chromosomes** 1
Sachihiro Matsunaga

2 **Mitochondria** 25
Shigeyuki Kawano

3 **Chloroplasts** 45
Shigeyuki Kawano

4 **The Endoplasmic Reticulum, Golgi Apparatuses,
and Endocytic Organelles** 71
Tetsuko Noguchi, Sachihiro Matsunaga, and Yasuko Hayashi

5 **Vacuoles and Storage Organelles** 89
Tetsuko Noguchi and Yasuko Hayashi

6 **Cytoskeletons** 107
Ichirou Karahara

7 **Cell Walls** 137
Tetsuko Noguchi

8 **Generative Cells** 157
Atsushi Sakai

9 **Meristems** 187
Hirokazu Tsukaya

Plates

1 Nuclei and Chromosomes 1

    1.1 Ultrastructural appearance of nuclei at different cell stages in high pressure frozen onion epidermal cells 2
    Ichirou Karahara, Lucas Andrew Staehelin, and Yoshinobu Mineyuki

    1.2 Morphology of nucleoli in tobacco BY-2 cultured cells 4
    Junko Hasegawa and Sachihiro Matsunaga

    1.3 Nuclear lamina localized at the nuclear periphery in interphase and at chromosomes in mitotic phase 6
    Yuki Sakamoto and Shingo Takagi

    1.4 Nuclei of multinucleate cells in *Hydrodictyon reticulatum* 8
    Manabu Tanaka and Kyoko Hatano

    1.5 Meiotic chromosomes of *Arabidopsis thaliana* pollen mother cells 10
    Yoshitaka Azumi

    1.6 Multicolor FISH of *Pinus* chromosomes 12
    Masahiro Hizume and Fukashi Shibata

    1.7 Chromosome painting and FISH of distal end satellite DNAs in dioecious plants with sex chromosomes 14
    Fukashi Shibata, Yusuke Kazama, Shigeyuki Kawano and Masahiro Hizume

    1.8 Kinetochore and microtubule dynamics during cell division of tobacco BY-2 cells visualized by live-cell imaging 16
    Daisuke Kurihara and Sachihiro Matsunaga

    1.9 Visualization of chromatin dynamics in the root of *Arabidopsis thaliana* 18
    Takeshi Hirakawa and Sachihiro Matsunaga

    1.10 Specific contribution of condensin II to sister centromere resolution in *Cyanidioschyzon merolae* 20
    Takayuki Fujiwara and Tatsuya Hirano

    1.11 Endomitosis induces a giant polyploid cell on the leaf epidermis 22
    Sachihiro Matsunaga and Masaki Ito

2 Mitochondria 25

    2.1 Mechanisms of division and inheritance of mitochondria and chloroplasts 26
    Tsuneyoshi Kuroiwa and Isamu Miyakawa

    2.2 Mitochondrial nucleoid of *Physarum polycephalum* 28
    Narie Sasaki

    2.3 Uniparental inheritance of mitochondria during mating of *Didymium iridis* 30
    Yohsuke Moriyama and Shigeyuki Kawano

2.4 Giant mitochondrion in synchronized *Chlamydomonas* cells 32
Tomoko Ehara and Tetsuaki Osafune

2.5 Dynamic transition of mitochondrial morphologies
during germination in living zygospore 34
Hiroaki Aoyama and Soichi Nakamura

2.6 Mitochondrial nucleoids in the *Euglena gracilis*
mitochondrial network 36
Yasuko Hayashi and Katsumi Ueda

2.7 Mitochondrial fission and fusion in an onion epidermal cell 38
Shin-ichi Arimura

2.8 Mitochondria in *Arabidopsis* guard cells 40
Chieko Saito

2.9 Mitochondria of thermogenic skunk cabbage 42
Mayuko Sato and Yasuko Ito-Inaba

3 Chloroplasts 45

3.1 Chloroplast division by the plastid-dividing ring 46
Shin-ya Miyagishima and Tsuneyoshi Kuroiwa

3.2 Chloroplasts divide by contraction of a bundle
of polyglucan nanofilaments 48
Yamato Yoshida, Haruko Kuroiwa, and Tsuneyoshi Kuroiwa

3.3 Cyanelle division of the glaucocystophyte
alga *Cyanophora paradoxa* 50
Haruki Hashimoto, Mayuko Sato, and Shigeyuki Kawano

3.4 3D distribution of RuBisCO in synchronized *Euglena* cells 52
Tetsuaki Osafune, Tomoko Ehara, and Shuji Sumida

3.5 Developing and degenerating chloroplasts
in *Haematococcus pluvialis* 54
Shuhei Ota and Shigeyuki Kawano

3.6 Monoplastidic cells in lower land plants 56
Masaki Shimamura

3.7 Dimorphic chloroplasts in the epidermis
of the aquatic angiosperm *Podostemaceae* family 58
Rieko Fujinami

3.8 Distribution of chloroplasts and mitochondria
in *Kalanchoë blossfeldiana* mesophyll cells 60
Ayumu Kondo

3.9 Etioplast prolamellar bodies in *Arabidopsis thaliana*
etiolated cotyledon 62
Yasuko Hayashi

3.10 Chloroplasts and mitochondria in *Sorghum*
bundle sheath cells 64
Chieko Saito, Yoshihiro Kobae, and Takashi Sazuka

3.11 Chloroplast division machinery in *Pelargonium zonale* 66
Haruko Kuroiwa and Tsuneyoshi Kuroiwa

3.12 Active digestion of paternal chloroplast DNA
in a young *Chlamydomonas reinhardtii* zygote 68
Yoshiki Nishimura

**4 The Endoplasmic Reticulum, Golgi Apparatuses,
and Endocytic Organelles** 71

4.1 Endoplasmic reticulum throughout the cytoplasm 72
Haruko Ueda, Etsuo Yokota, and Ikuko Hara-Nishimura

4.2 Endoplasmic reticulum in the green alga *Botryococcus braunii* 74
Tetsuko Noguchi

	4.3	**ER body in cotyledon epidermal cells**	76
		Yasuko Hayashi and Toshiyuki Sakurai	
	4.4	**Golgi apparatuses in a *Brachypodium* root cap peripheral cell**	78
		Mayuko Sato	
	4.5	**Golgi bodies in mature pollen of *Tradescantia reflexa***	80
		Tetsuko Noguchi	
	4.6	**Golgi bodies and the *trans*-Golgi networks in *Botryococcus braunii***	82
		Tetsuko Noguchi	
	4.7	**Clathrin-coated buds and vesicles in *Botryococcus braunii***	84
		Tetsuko Noguchi	
	4.8	**Spatio-temporal dynamics of endocytic vesicle formation in *Arabidopsis thaliana***	86
		Masaru Fujimoto and Takashi Ueda	
5	**Vacuoles and Storage Organelles**		89
	5.1	**Central vacuole in *Arabidopsis thaliana* pistil cells**	90
		Tetsuko Noguchi	
	5.2	**Vacuoles under salt stress**	92
		Tetsuro Mimura and Kohei Hamaji	
	5.3	**Autophagosomes and autolysosomes in plant cells**	94
		Yuko Inoue and Yuji Moriyasu	
	5.4	**Transition of peroxisomes from glyoxysomes to leaf peroxisomes during greening in cotyledon**	96
		Yasuko Hayashi and Shoji Mano	
	5.5	**Dynamics of embryonic pea leaf cells during early germination**	98
		Yasuko Kaneko	
	5.6	**Lipids and astaxanthin are major contents of subcellular changes during encystment in *Haematococcus pluvialis***	100
		Shuhei Ota and Shigeyuki Kawano	
	5.7	**Lipid accumulation in the green alga *Botryococcus braunii***	102
		Reiko Suzuki and Tetsuko Noguchi	
	5.8	**Production of oil bodies in response to nitrogen starvation in *Chlamydomonas reinhardtii***	104
		Haruko Kuroiwa and Tsuneyoshi Kuroiwa	
6	**Cytoskeletons**		107
	6.1	**Microtubule systems in the cell cycle of onion root tips visualized in 3D**	108
		Yoshinobu Mineyuki	
	6.2	**Microtubule-dependent microtubule nucleation in a tobacco BY-2 cell**	110
		Takashi Murata	
	6.3	**Ultrastructural appearance of microtubules in high-pressure frozen onion epidermal cells**	112
		Ichirou Karahara, Takashi Murata, Lucas Andrew Staehelin, and Yoshinobu Mineyuki	
	6.4	**Microtubules and their end structures in high-pressure frozen onion epidermal cells visualized by electron tomography**	114
		Ichirou Karahara and Yoshinobu Mineyuki	
	6.5	**Microtubule organizing centers in bryophytes**	116
		Masaki Shimamura and Yoshinobu Mineyuki	
	6.6	**Selective disappearance of female centrioles after fertilization in brown algae**	118
		Chikako Nagasato and Taizo Motomura	

6.7 Spindle formation in brown algae 120
Chikako Nagasato

6.8 Helical rows of microtubules in *Euglena* pellicles 122
Tetsuko Noguchi

6.9 Spindle pole body during meiosis I in the budding
yeast *Saccharomyces cerevisiae* 124
Aiko Hirata and Shigeyuki Kawano

6.10 Actin filaments in *Lilium longiflorum* pollen protoplasts 126
Ichiro Tanaka

6.11 Dynamics of actin filaments in the liverwort,
Marchantia polymorpha 128
Atsuko Era and Takashi Ueda

6.12 Two actin structures in dormant *Dictyostelium discoideum* spores 130
Masazumi Sameshima

6.13 Actin-microtubule interaction during preprophase band
formation in onion root tips visualized
by immuno-fluorescence microscopy 132
Miyuki Takeuchi and Yoshinobu Mineyuki

6.14 Microtubules direct the layered structure of angiosperm
shoot apical meristems (SAMs) 134
Shuichi Sakaguchi

7 Cell Walls 137

7.1 Ribbon-like fibrillar network of glucan in reverting
Schizosaccharomyces pombe protoplast 138
Masako Osumi

7.2 Mother and daughter cell walls during autosporulation
in the green alga *Chlorella vulgaris* 140
Maki Yamamoto and Shigeyuki Kawano

7.3 Great-grandmother, grandmother, mother, and daughter cell
walls during budding in the green alga *Marvania geminata* 142
Maki Yamamoto, Satomi Owari, and Shigeyuki Kawano

7.4 Formation of amphiesmal vesicles and thecal plates
in the dinoflagellate *Scrippsiella hexapraecingula* 144
Satoko Sekida and Kazuo Okuda

7.5 The elaborate shape of *Micrasterias* is formed by a primary
cell wall containing pectin 146
Tetsuko Noguchi

7.6 Cellulose-synthesizing rosettes in the green algae
Micrasterias and *Closterium* 148
Tetsuko Noguchi

7.7 Localization of typical cell wall polysaccharides pectin
and β-1,3/1,4 mixed linkage glucan in *Arabidopsis thaliana*
and *Oryza sativa* 150
Ryusuke Yokoyama, Hideki Narukawa, and Kazuhiko Nishitani

7.8 Plasmodesmata directly connect the cytoplasm of neighboring
plant cells 152
Yasuko Hayashi

7.9 Meshwork structure of the Casparian strip 154
Yoshihiro Honma and Ichirou Karahara

8 Generative Cells 157

8.1 **A mating-pair of seaweed *Ulva compressa* gametes with asymmetrical mating structure positions** 158
Yuko Mogi and Shigeyuki Kawano

8.2 **Spermatogenesis in *Marchantia polymorpha*** 160
Katsumi Ueda and Tetsuko Noguchi

8.3 **Motile sperms released in the ovule of an extinct Permian gymnosperm *Glossopteris*** 162
Harufumi Nishida

8.4 **Multiflagellated sperm of *Ginkgo biloba* L.** 164
Shinichi Miyamura

8.5 **Pollen exine and male gametic nucleus of *Lilium longiflorum*** 166
Norifumi Mogami and Ichiro Tanaka

8.6 **Developing *Arabidopsis* pollen grain containing a young generative cell with some mitochondria and no plastids** 168
Chieko Saito and Keiko Shoda

8.7 **The selective increase or decrease of organelle DNAs in young generative cells controls cytoplasmic inheritance in higher plants** 170
Noriko Nagata

8.8 **Dimorphic *Plumbago auriculata* sperm cells** 172
Chieko Saito

8.9 **Pollen tube guidance toward the ovule** 174
Masahiro M. Kanaoka

8.10 **Semi-in vitro *Torenia* system for live-cell analysis of plant fertilization** 176
Tetsuya Higashiyama

8.11 **Protoplasts from plant female gametophytes** 178
Masahiro M. Kanaoka

8.12 **Egg cells with giant mitochondria in a higher plant, *Pelargonium zonale* Ait** 180
Haruko Kuroiwa and Tsuneyoshi Kuroiwa

8.13 **Zygote and sperm cells during early embryogenesis in a higher plant, *Pelargonium zonale* Ait** 182
Haruko Kuroiwa and Tsuneyoshi Kuroiwa

8.14 **Cell geometry in a whole *Arabidopsis* seed visualized by X-ray micro-CT** 184
Yoshinobu Mineyuki, Aki Fukuda, Daisuke Yamauchi, and Ichirou Karahara

9 Meristems 187

9.1 **Two types of meristem involved in development of the fern gametophyte** 188
Ryoko Imaichi

9.2 **Structures of fern and lycophyte shoot apical meristems (SAMs)** 190
Ryoko Imaichi

9.3 **Structures of angiosperm SAMs** 192
Ryoko Imaichi

9.4 ***Arabidopsis thaliana* leaf blade and leaf petiole** 194
Yasunori Ichihashi, Kensuke Kawade, and Hirokazu Tsukaya

9.5 **Structures of fern and lycophyte root apical meristems (RAMs)** 196
Ryoko Imaichi

9.6 **Structures of angiosperm RAMs** 198
Ryoko Imaichi

9.7 **Asymmetric cell division forms endodermis and cortex
in *Arabidopsis thaliana* root** 200
Mai Takagi and Sachihiro Matsunaga

Contributors

Hiroaki Aoyama Center of Molecular Biosciences, Tropical Biosphere Research Center University of the Ryukyus, Okinawa, Japan

Laboratory of Cell and Functional Biology, Faculty of Science, University of the Ryukyus, Okinawa, Japan

(Plate 2.5)

Shin-ichi Arimura Graduate School of Agricultural and Life Sciences, The University of Tokyo, Tokyo, Japan

(Plate 2.7)

Yoshitaka Azumi Department of Biological Sciences, Faculty of Science Kanagawa University, Kanagawa, Japan

(Plate 1.5)

Tomoko Ehara Department of Microbiology, Tokyo Medical University, Tokyo, Japan

(Plates 2.4 and 3.4)

Atsuko Era Department of Biological Sciences, Graduate School of Science, The University of Tokyo, Tokyo, Japan

(Plate 6.11)

Masaru Fujimoto Department of Agricultural and Environmental Biology, Graduate School of Agricultural and Life Sciences, The University of Tokyo, Tokyo, Japan

(Plate 4.8)

Rieko Fujinami Department of Chemical and Biological Sciences, Japan Women's University, Tokyo, Japan

(Plate 3.7)

Takayuki Fujiwara Chromosome Dynamics Laboratory RIKEN, Saitama, Japan

(Plate 1.10)

Aki Fukuda Department of Life Science, Faculty of Science University of Hyogo, Hyogo, Japan

(Plate 8.14)

Kohei Hamaji Department of Biology, Graduate School of Science, Kobe University, Kobe, Japan

(Plate 5.2)

Ikuko Hara-Nishimura Department of Botany, Graduate School of Science, Kyoto University, Kyoto, Japan

(Plate 4.1)

Junko Hasegawa Department of Applied Biological Science, Faculty of Science and Technology Tokyo University of Science, Chiba, Japan

(Plate 1.2)

Haruki Hashimoto Department of Biological Sciences, Faculty of Science, Kanagawa University, Hiratsuka, Japan
(Plate 3.3)

Kyoko Hatano Department of Interdisciplinary Environment, Graduate School of Human and Environmental Studies, Kyoto University, Kyoto, Japan
(Plate 1.4)

Yasuko Hayashi Department of Environmental Science, Graduate School of Science and Technology, Niigata University, Niigata, Japan
(Plates 2.6, 3.9, 4.3, 5.4 and 7.8)

Tetsuya Higashiyama Division of Biological Science, Institute of Transformative Bio-Molecules (WPI-ITbM), Graduate School of Science/JST ERATO Higashiyama Live-Holonics Project, Nagoya University, Aichi, Japan
(Plate 8.10)

Takeshi Hirakawa Department of Applied Biological Science, Faculty of Science and Technology Tokyo University of Science, Chiba, Japan
(Plate 1.9)

Tatsuya Hirano Chromosome Dynamics Laboratory RIKEN, Saitama, Japan
(Plate 1.10)

Aiko Hirata Bioimaging Center of Integrated Biosciences, Graduate School of Frontier Sciences, The University of Tokyo, Chiba, Japan
(Plate 6.9)

Masahiro Hizume Faculty of Education, Ehime University, Matsuyama, Japan
(Plates 1.6 and 1.7)

Yoshihiro Honma Department of Biology, Graduate School of Science and Engineering, University of Toyama, Toyama, Japan
(Plate 7.9)

Yasunori Ichihashi Department of Plant Biology, University of California at Davis, Davis, CA, USA
(Plate 9.4)

Ryoko Imaichi Department of Chemical and Biological Sciences, Japan Women's University, Tokyo, Japan
(Plates 9.1, 9.2, 9.3, 9.5, and 9.6)

Yuko Inoue Department of Regulatory Biology, Graduate School of Science and Engineering, Saitama University, Saitama, Japan
(Plate 5.3)

Masaki Ito Graduate School of Bioagricultural Sciences, Nagoya University, Nagoya, Japan
(Plate 1.11)

Yasuko Ito-Inaba Organization for Promotion of Tenure Track, University of Miyazaki, Miyazaki, Japan
(Plate 2.9)

Masahiro M. Kanaoka Division of Biological Science, Graduate School of Science, Nagoya University, Aichi, Japan
(Plates 8.9 and 8.11)

Yasuko Kaneko Biology Section in the Faculty of Education, Saitama University, Saitama, Japan
(Plate 5.5)

Ichirou Karahara Department of Biology, Graduate School of Science and Engineering, University of Toyama, Toyama, Japan
(Plates 1.1, 6.3, 6.4, 7.9, and 8.14)

Kensuke Kawade RIKEN CSRS, Kanagawa, Japan
(Plate 9.4)

Shigeyuki Kawano Department of Integrated Biosciences Graduate School of Frontier Sciences, The University of Tokyo, Chiba, Japan
(Plates 1.7, 2.3, 3.3, 3.5, 5.6, 6.9, 7.2, 7.3, and 8.1)

Yusuke Kazama Ion Beam Breeding Team RIKEN Innovation Center, Saitama, Japan
(Plate 1.7)

Yoshihiro Kobae Faculty of Agriculture, Graduate School of Agricultural and Life Sciences, The University of Tokyo, Tokyo, Japan
(Plate 3.10)

Ayumu Kondo Faculty of Agriculture, Meijo University, Nagoya, Japan
(Plate 3.8)

Daisuke Kurihara Division of Biological Science, Graduate School of Science, Nagoya University, Aichi, Japan
(Plate 1.8)

Tsuneyoshi Kuroiwa CREST, Initiative Research Unit, College of Science, Rikkyo University, Tokyo, Japan
(Plates 2.1, 3.1, 3.2, 3.11, 5.8, 8.12, and 8.13)

Haruko Kuroiwa CREST, Initiative Research Unit, College of Science, Rikkyo University, Tokyo, Japan
(Plates 3.2, 3.11, 5.8, 8.12, and 8.13)

Shoji Mano Department of Cell Biology, National Institute for Basic Biology, Okazaki, Japan
(Plate 5.4)

Sachihiro Matsunaga Department of Applied Biological Science, Faculty of Science and Technology Tokyo University of Science, Chiba, Japan
(Plates 1.2, 1.8, 1.9, 1.11, and 9.7)

Tetsuro Mimura Department of Biology, Graduate School of Science, Kobe University, Kobe, Japan
(Plate 5.2)

Yoshinobu Mineyuki Department of Life Science, Graduate School of Life Science, University of Hyogo, Hyogo, Japan
(Plates 1.1, 6.1, 6.3, 6.4, 6.5, 6.13, and 8.14)

Shin-ya Miyagishima Center for Frontier Research, National Institute of Genetics, Shizuoka, Japan
(Plate 3.1)

Isamu Miyakawa Faculty of Science, Yamaguchi University, Yamaguchi, Japan
(Plate 2.1)

Shinichi Miyamura Faculty of Life and Environmental Sciences, University of Tsukuba, Ibaraki, Japan
(Plate 8.4)

Norifumi Mogami Department of Biological and Chemical Systems, Engineering Kumamoto National College of Technology, Kumamoto, Japan
(Plate 8.5)

Yuko Mogi Department of Integrated Biosciences, Graduate School of Frontier Sciences, The University of Tokyo, Chiba, Japan
(Plate 8.1)

Yohsuke Moriyama Department Anatomy II and Cell Biology, School of Medicine, Fujita Health University, Aichi, Japan
(Plate 2.3)

Yuji Moriyasu Department of Regulatory Biology, Graduate School of Science and Engineering, Saitama University, Saitama, Japan
(Plate 5.3)

Taizo Motomura Muroran Marine Station, Field Science Center for Northern Biosphere, Hokkaido University, Hokkaido, Japan
(Plate 6.6)

Takashi Murata Division of Evolutionary Biology, National Institute for Basic Biology, Aichi, Japan
(Plates 6.2 and 6.3)

Chikako Nagasato Muroran Marine Station, Field Science Center for Northern Biosphere, Hokkaido University, Hokkaido, Japan
(Plates 6.6 and 6.7)

Noriko Nagata Faculty of Science, Japan Women's University, Tokyo, Japan
(Plate 8.7)

Soichi Nakamura Laboratory of Cell and Functional Biology, Faculty of Science University of the Ryukyus, Okinawa, Japan
(Plate 2.5)

Hideki Narukawa Laboratory of Plant Cell Wall Biology, Graduate School of Life Sciences, Tohoku University, Miyagi, Japan
(Plate 7.7)

Harufumi Nishida Department of Biological Sciences, Faculty of Science and Engineering Chuo University, Tokyo, Japan
(Plate 8.3)

Yoshiki Nishimura Department of Botany, Kyoto University, Kyoto, Japan
(Plate 3.12)

Kazuhiko Nishitani Laboratory of Plant Cell Wall Biology, Graduate School of Life Sciences, Tohoku University, Miyagi, Japan
(Plate 7.7)

Tetsuko Noguchi Course of Biological Sciences, Faculty of Science Nara Women's University, Nara, Japan
(Plates 4.2, 4.5, 4.6, 4.7, 5.1, 5.7, 6.8, 7.5, 7.6, and 8.2)

Kazuo Okuda Graduate School of Kuroshio Science, Kochi University, Kochi, Japan
(Plate 7.4)

Tetsuaki Osafune Beppubay Research Institute for Applied Microbiology, Kitsuki, Japan
(Plates 2.4 and 3.4)

Masako Osumi Integrated Imaging Research Support, Tokyo, Japan
(Plate 7.1)

Shuhei Ota Department of Integrated Biosciences, Graduate School of Frontier Sciences, The University of Tokyo, Chiba, Japan
(Plates 3.5 and 5.6)

Satomi Owari Neo-Morgan Laboratory Incorporated Research & Development, Biotechnology Research Center, Kanagawa, Japan
(Plate 7.3)

Chieko Saito Department of Biological Sciences, Graduate School of Science, The University of Tokyo, Tokyo, Japan
(Plates 2.8, 3.10, 8.6, and 8.8)

Shuichi Sakaguchi Course of Biological Sciences, Faculty of Science Nara Women's University, Nara, Japan
(Plate 6.14)

Yuki Sakamoto Department of Biological Science, Graduate School of Science, Osaka University, Osaka, Japan
(Plate 1.3)

Toshiyuki Sakurai Department of Environmental Science, Graduate School of Science and Technology, Niigata University, Niigata, Japan
(Plate 4.3)

Masazumi Sameshima Integrated Imaging Research Support, Tokyo, Japan
(Plate 6.12)

Narie Sasaki Division of Biological Science, Graduate School of Science, Nagoya University, Aichi, Japan
(Plate 2.2)

Mayuko Sato RIKEN Center for Sustainable Resource Science, Kanagawa, Japan
(Plates 2.9, 3.3, and 4.4)

Takashi Sazuka Bioscience and Biotechnology Center, Nagoya University, Nagoya, Japan
(Plate 3.10)

Satoko Sekida Graduate School of Kuroshio Science, Kochi University, Kochi, Japan
(Plate 7.4)

Fukashi Shibata Institute of Plant Science and Resources, Okayama University, Kurashiki, Japan
(Plates 1.6 and 1.7)

Masaki Shimamura Department of Biological Science, Faculty of Science Hiroshima University, Hiroshima, Japan
(Plates 3.6 and 6.5)

Keiko Shoda Laboratory for Cell Function Dynamics, RIKEN Brain Science Institute, Saitama, Japan
(Plate 8.6)

Lucas Andrew Staehelin MCD Biology, University of Colorado, Boulder, CO, USA
(Plates 1.1 and 6.3)

Shuji Sumida Department of Microbiology, Tokyo Medical University, Tokyo, Japan
(Plate 3.4)

Reiko Suzuki Course of Biological Sciences, Faculty of Science Nara Women's University, Nara, Japan
(Plate 5.7)

Shingo Takagi Department of Biological Science, Graduate School of Science, Osaka University, Osaka, Japan
(Plate 1.3)

Mai Takagi Department of Applied Biological Science, Faculty of Science and Technology Tokyo University of Science, Chiba, Japan
(Plate 9.7)

Miyuki Takeuchi Department of Biomaterial Sciences, Graduate School of Agricultural and Life Sciences, The University of Tokyo, Tokyo, Japan
(Plate 6.13)

Manabu Tanaka Department of Interdisciplinary Environment, Graduate School of Human and Environmental Studies, Kyoto University, Kyoto, Japan
(Plate 1.4)

Ichiro Tanaka Graduate School of Nanobioscience, Yokohama City University, Kanagawa, Japan
(Plates 6.10 and 8.5)

Hirokazu Tsukaya Department of Biological Sciences, Graduate School of Science, The University of Tokyo, Tokyo, Japan
(Plate 9.4)

Haruko Ueda Department of Botany Graduate School of Science, Kyoto University, Kyoto, Japan
(Plate 4.1)

Takashi Ueda Department of Biological Sciences, Graduate School of Science, The University of Tokyo, Tokyo, Japan
(Plates 4.8 and 6.11)

Katsumi Ueda
(Plates 2.6 and 8.2)

Maki Yamamoto Institute of Natural Sciences, Senshu University, Kanagawa, Japan
(Plates 7.2 and 7.3)

Daisuke Yamauchi Department of Life Science, Graduate School of Life Science, University of Hyogo, Hyogo, Japan
(Plate 8.14)

Etsuo Yokota Department of Life Science, Graduate School of Life Science, University of Hyogo, Hyogo, Japan
(Plate 4.1)

Ryusuke Yokoyama Laboratory of Plant Cell Wall Biology Graduate School of Life Sciences, Tohoku University, Miyagi, Japan
(Plate 7.7)

Yamato Yoshida CREST, Initiative Research Unit, College of Science, Rikkyo University, Tokyo, Japan

Department of Plant Biology Michigan State University, East Lansing, MI, USA
(Plate 3.2)

Abbreviations

ADZ	Actin-depleted zone
5-AU	5-Aminouracil
bp	*Biparental*
CAM	Crassulacean-acid-metabolism
CCD	Charge coupled device
cMT	Cortical microtubule
CT	Computed tomography
DAPI	4',6-Diamidino-2-phenylindole
DASMPI	2-(p-Dimethylamino-styryl)-1-methylpyridine iodine
$DiOC_6(3)$	3,3'-Dehexyloxacarbocyanine iodide
DRP	Dynamin-related protein
EdU	5-Ethynyl-2'-deoxyuridine
ER	Endoplasmic reticulum
FE-SEM	Field-emission scanning electron microscope
FISH	Fluorescence in situ hybridization
FITC	Fluorescein isothiocyanate
GFP	Green fluorescent protein
HPF	High-pressure freezing
ICM	Interphase cortical microtubule
LVSEM	Low voltage scanning electron microscope
MeJA	Methyl jasmonate
mKO	Monomeric Kusabira Orange
MLG	Mixed linkage glucans
mPS-PI	Modified pseudo-Schiff propidium iodide
MS	Murashige and Skoog
MT	Microtubule
MTOC	Microtubule organizing center
MVB	Multivesicular body
OB	Oil body
PAR	Pseudoautosomal region
PD ring	Plastid-dividing ring
PI	Propidium iodide
PM	Plasma membrane
PP	Phragmoplast
PPB	Preprophase band
PTS1	Peroxisomal targeting signal 1
RAM	Root apical meristem
rER	Rough endoplasmic reticulum
RFP	Red fluorescent protein
RMS	Radial microtubule system
RuBisCO	Ribulose-1,5-bisphosphate carboxylase/oxygenase
SAHH	S-adenosyl-L-homocysteine hydrolase

SAM	Shoot apical meristem
SEM	Scanning electron microscope
sER	Smooth endoplasmic reticulum
SPB	Spindle pole body
TAG	Triacylglycerol
TEM	Transmission electron microscope
TGN	*trans*-Golgi network
TIRFM	Total internal reflection fluorescence microscopy
VIAFM	Variable incidence angle fluorescence microscopy
VIMPCS	Video-intensified microscopic photon counting system

Marks in Plates

AF	Actin filament
Am	Amyloplast
cCh	Clumped chloroplasts
CCP	Clathrin-coated pit
CCV	Clathrin-coated vesicle
Ce	Centriol
Ch	Chloroplast
Ch^{-L}	Large chloroplast
Ch^{-S}	Small chloroplast
Col	Collenchyma
Cor	Cortex
Cot	Cotyledon
CW	Cell wall
E	Eyespot
Ep	Epidermal cell
ER	Endoplasmic reticulum
G	Golgi body
GC	Generative cell
Hy	Hypocotyl
If	Interfascicular fiber
L	Lysosome
M	Mitochondrion
Me	Mesophyll
MT	Microtubule
MVB	Multi vesicular body
N	Nucleus
N^{-GC}	Generative cell nucleus
N^{-S}	Sperm cell nucleus
N^{-Z}	Zygote nucleus
OB	Oil body
P	Pyrenoid
Pa	Parenchyma
Ph	Phloem
PM	Plasma membrane
PT	Pollen tube
Ra	Radicle
SG	Starch grain
TGN	*trans*-Golgi network
V	Vacuole
VC	Vegetative cell
X	X chromosome
Y	Y chromosome

Nuclei and Chromosomes

Sachihiro Matsunaga

The nucleus is a double membrane-enclosed organelle in which the genome is packaged. DNA replication and RNA transcription occur in the nucleus. Franz Bauer first described the nucleus in orchid cells as observed under a microscope in 1802. Robert Brown observed details of the nucleus in orchid cells and named the organelle "nucleus" in 1831.

In this chapter, I. Karahara et al. reveal the ultrastructure of onion nuclei using a high pressure freezing technique. The cross section shows heterochromatic regions of interphase nuclei and mitotic chromosomes as dark staining regions of the chromosomes. The image of interphase nuclei demonstrates that plant nuclei with a large genome have globular domains in the nucleoplasm.

The nucleus contains various subnuclear structures. The most conspicuous structure in the nucleus is the nucleolus. At interphase, the nucleolus is responsible for ribosomal RNA synthesis and ribosome assembly. J. Hasegawa and S. Matsunaga present dynamic changes of nucleoli in tobacco BY-2 cultured cells. Nucleolus morphology changes dynamically through the cell cycle. The outermost structure of the nucleus is the nuclear membrane, which consists of inner and outer membranes. The inner membrane is backed by lamina. Y. Sakamoto and S. Takagi demonstrate the localization of a lamina component LITTLE NUCLEI 1 (CROWDED NUCLEI1). In addition to being present in animal skeletal muscle cells, multinucleated cells are also found in plants. M. Tanaka and K. Hatano present multinucleated cells in a green alga.

In the nucleus, DNA is packaged into chromatin, which is condensed to form chromosomes. Chromosomes were first described in the nucleus by Anton Schneider in 1873, and plant chromosomes were first described by Eduard Strasburger shortly thereafter in 1875. Condensed chromosomes are clearly observed during mitosis and meiosis. Y. Azumi presents the morphology of chromosomes at each stage of meiosis in male gametogenesis. Fluorescent in situ hybridization (FISH) has been a powerful technique for analyses of distribution of DNA sequences and chromosome organization. M. Hizume and F. Shibata demonstrate FISH of pine mitotic chromosomes. The behavior of chromosomes determines heredity, and chromosomal rearrangements are a driving force in evolution. Several sex chromosomes in dioecious plants genetically determine sex. F. Shibata et al. demonstrate the sex chromosomes in *Rumex acetosa* and *Silene latifolia*.

The accurate distribution of the genome during mitosis and meiosis is a direct consequence of chromosome dynamics. Chromosome alignment and segregation is regulated by the interaction of kinetochores and microtubules. D. Kurihara and S. Matsunaga perform live cell imaging to reveal the dynamics of kinetochores and microtubules during mitosis. Recently, chromatin movement was reported in interphase nuclei, suggesting that chromatin is not stable but structurally dynamic. T. Hirakawa and S. Matsunaga visualize chromatin in the root. Condensin is a regulatory protein for chromosome condensation. T. Fujiwara and T. Hirano show the distribution of condensin in primitive red alga.

The development processes of plants are often accompanied by endopolyploidy, which mainly arises from endoreduplication or endomitosis. In endoreduplication, also known as endoreplication or endocycle, DNA replication during the S phase is not followed by subsequent mitosis, leading to a polyploid cell. By contrast, endomitosis lacks sister chromatid segregation and cytokinesis, similarly resulting in a doubling of ploidy. In both cases, the intranuclear DNA content doubles with every cell cycle, giving rise to cells with high DNA content. S. Matsunaga and M. Ito demonstrate that endomitosis and endoreduplication forms gigas and pavement cells on leaf epidermis, respectively.

S. Matsunaga (✉)
Department of Applied Biological Science, Faculty of Science and Technology, Tokyo University of Science, 2641 Yamazaki, Noda, Chiba 278-8510, Japan
e-mail: sachi@rs.tus.ac.jp

T. Noguchi et al. (eds.), *Atlas of Plant Cell Structure*,
DOI 10.1007/978-4-431-54941-3_1, © Springer Japan 2014

Plate 1.1

Ultrastructural appearance of nuclei at different cell stages in high pressure frozen onion epidermal cells

The basal region of young onion (*Allium cepa* L. cv. Highgold Nigou) cotyledons is an excellent system for studying plant cell mitosis because ~3 % of the epidermal cells are undergoing mitosis at any time. Such cells were used to study ultrastructural membrane and cytoskeletal changes during formation of the preprophase band, which defines the future site of cell division [1].

Electron micrographs show ultrathin sections of onion epidermal cells exhibiting exceptionally well-preserved nuclei at different stages of mitosis. Cross sections of an interphase (**A**) and a late prophase cell (**B**) (adapted from [1]). Note the differences in staining patterns of the dark, condensed chromosome regions. A longitudinal section of a cluster of epidermal cells at different stages of mitosis after synchronization with the thymine analogue 5-aminouracil (5-AU) (**C**). Inter, interphase; Late pro, late prophase; Prometa, prometaphase; Meta, metaphase; Ana, anaphase.

To avoid chemical fixation artifacts, high-pressure freezing (HPF), by which entire tissues are immobilized in ~1 ms, was employed. 1–2 mm long basal cotyledon segments of 3 day old onion seedlings were cut with a razor blade while submerged in 0.1 M sucrose, and immediately frozen in a BAL-TEC HPM 010 high-pressure freezer (Boeckeler). Frozen samples were freeze-substituted in 2 % (w/v) OsO_4 in anhydrous acetone and then embedded in Spurr resin. Ultrathin sections were stained with uranyl acetate and Reynold's lead citrate, and were imaged with a transmission electron microscope. Mitotic synchronization involved treating 2.5 day old onion cotyledons with a 0.6 mg/mL solution of 5-AU for 12 h then transferring the tissue to drug-free medium for 8–9 h followed by high pressure freezing. Scale bars: 5 nm (**A**), 10 nm (**C**).

Contributors

Ichirou Karahara[1]*, Lucas Andrew Staehelin[2], Yoshinobu Mineyuki[3], [1]Department of Biology, Graduate School of Science and Engineering, University of Toyama, Toyama 930-8555, Japan, [2]MCD Biology, University of Colorado, Boulder, CO 80309-0347, USA, [3]Department of Life Science, Graduate School of Life Science, University of Hyogo, 2167 Shosha, Himeji, Hyogo 671-2280, Japan
*E-mail: karahara@sci.u-toyama.ac.jp

References

1. Karahara I, Suda J, Tahara H, Yokota E, Shimmen T, Misaki K, Yonemura S, Staehelin A Mineyuki Y (2009) The preprophase band is a localized center of clathrin-mediated endocytosis in late prophase cells of the onion cotyledon epidermis. Plant J 57:819–831. doi:10.1111/j.1365-313X.2008.03725.x

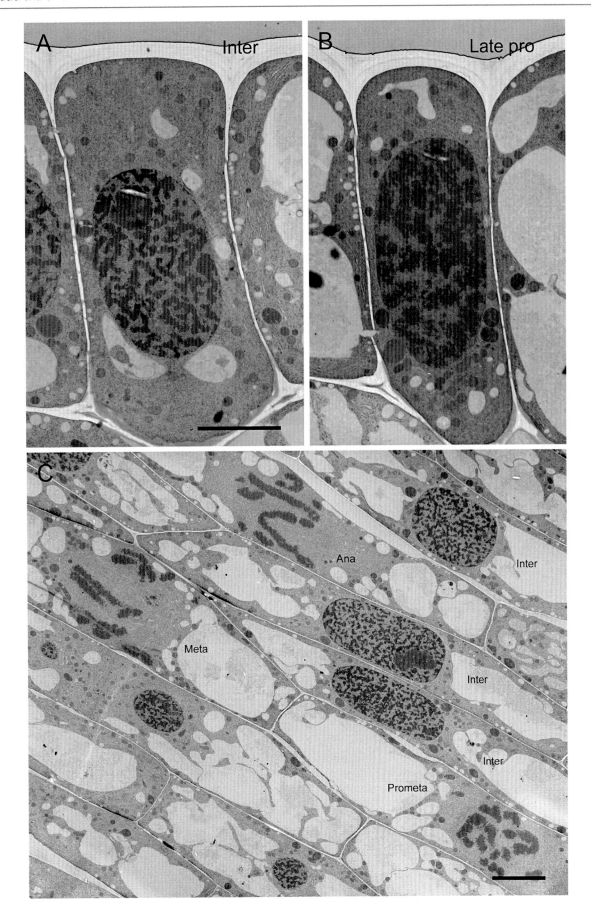

Plate 1.2

Morphology of nucleoli in tobacco BY-2 cultured cells

The nucleolus is a distinct subnuclear structure of eukaryotic cells. The biosynthesis of rRNA and ribosomes occurs in the nucleolus. Molecules required for the synthesis of rRNA and ribosomes such as rDNA, precursor rRNA, mature rRNA, RNA-processing enzymes, snoRNA, ribosomal proteins, and assembly proteins, aggregate at high density in the nucleolus. First, 35S rRNAs are transcribed by RNA polymerase I in the nucleolus. Then, rRNA spacer regions are removed to mature 18S, 5.8S, and 26S rRNA. Each rRNA is modified and subsequently packed into a precursor ribosome. Precursor ribosomes are carried from the nucleolus to the cytoplasm to become mature ribosomes. The size and number of nucleoli depend on the phase of the cell cycle [2]. A nucleolus resides in the nucleus until prophase and disappears from prometaphase to telophase. Many nucleolar proteins are localized on the peripheral region of mitotic chromosomes [3]. Proteins regulating ribosome synthesis are necessary for root growth and epidermal cell patterning in plants.

This figure represents nuclei, chromosomes and cell walls in blue and RNA in green of tobacco BY-2 culture cells (*Nicotiana tabacum* L. cv. Bright Yelleow-2). At prophase, chromatins begin to condensate into chromosomes, but the structure of nucleolus is remained (middle right). At metaphase, the nucleolus disappeared completely (lower left). At telophase (middle left) to early G1 phase (left), two to four nucleoli are formed in a nuclear. At late G1, S and G2 phases, nucleoli reassembled into one nucleolus in accordance with in the progression of G1 phase (center and left side).

A nucleolar-ID™ green detection kit (Cat. # 51009-500, Enzo) was used for visualization of nucleoli. The reagent is a cell-permeable nucleic acid stain that is selective for RNA. The reagent is essentially non-fluorescent in the absence of nucleic acids, but emits green fluorescence when bound to RNA. Because nucleoli contain abundant synthesized rRNA, they stain brightly. The cytoplasmic region is also stained. 1 μL of Nucleolar-IDTM green detection reagent was added to 500 μL of 2 day old BY-2 cells. The cells were incubated in the dark for 30 min at room temperature. After the addition of 1 mL of 1× Assay Buffer diluted with BY-2 medium, those were stained with 50 μg/mL 4′, 6-diamidino-2-phenylindole phenylindole (DAPI). Stained cells were observed under an upright microscope (BX53, Olympus) with a CCD camera (Cool Snap HQ2, Nippon Roper). Scale bar: 30 μm.

Contributors

Junko Hasegawa, Sachihiro Matsunaga*, Department of Applied Biological Science, Faculty of Science and Technology, Tokyo University of Science, 2641 Yamazaki, Noda, Chiba 278-8510, Japan
*E-mail: sachi@rs.tus.ac.jp

References

2. Matsunaga S, Katagiri Y, Nagashima Y, Sugiyama T, Hasegawa J, Hayashi K, Sakamoto T (2013) New insights into the dynamics of plant cell nuclei and chromosomes. Int Rev Cell Mol Biol 305:253–301. doi:10.1016/B978-0-12-407695-2.00006-8
3. Matsunaga S, Fukui K (2010) The chromosome peripheral proteins play an active role in chromosome dynamics. Biomol Concepts 1:157–164. doi:10.1515/bmc.2010.018

1 Nuclei and Chromosomes

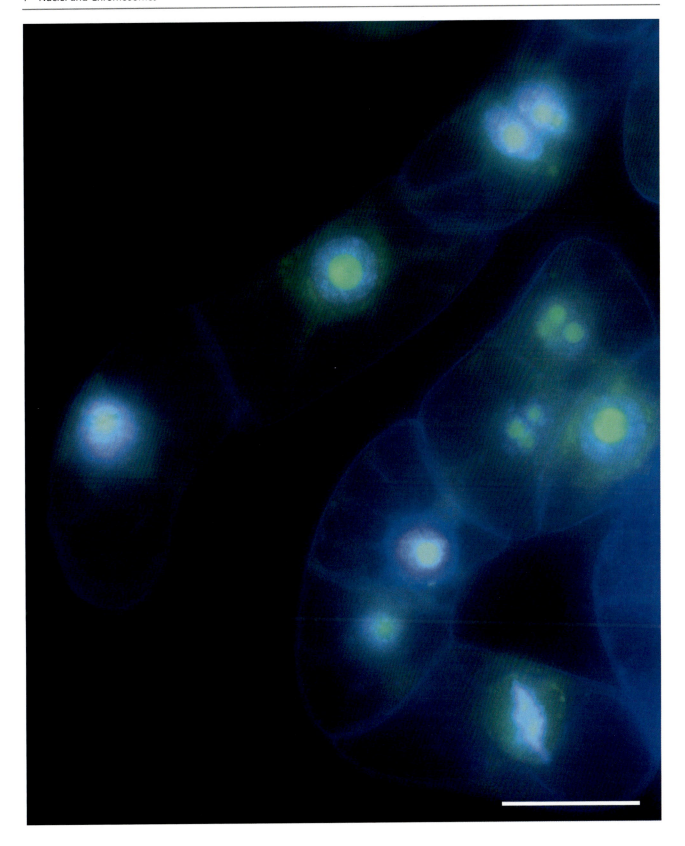

Plate 1.3

Nuclear lamina localized at the nuclear periphery in interphase and at chromosomes in mitotic phase

The nuclear lamina forms a meshwork beneath the inner nuclear envelope. Animal nuclear lamina, constructed of lamins and a variety of inner nuclear envelope proteins, contributes not only to an increase in mechanical toughness but also to regulation of chromatin distribution and gene expression. The components and biological significance of plant nuclear lamina are not nearly as well understood. The *Arabidopsis thaliana* LITTLE NUCLEI (LINC) (also called CROWDED NUCLEI, CRWN) family proteins were discovered based on homology to carrot nuclear matrix constituent protein [4] and detected in the nuclear lamina fraction by liquid chromatography tandem mass spectroscopy. The intracellular localization of LINC1-GFP was investigated in the fixed root apical meristem. The fluorescent images were adapted from [5]. LINC1 was localized mainly to the nuclear periphery in interphase cells (**A**), to the condensing chromosomes after prometaphase and into anaphase (**B**), transferred from decondensing chromatin to the reassembling nuclear envelope in early telophase (**C**), and then again to the nuclear periphery during late telophase (**D**). Hoechst signals are in magenta and GFP signals are in green.

Sample roots were fixed in 4 % (w/v) formaldehyde, freshly prepared from paraformaldehyde, in PIPES buffer (10 mM EGTA, 5 mM $MgSO_4$, and 50 mM PIPES at pH 7.0) for 1 h. Fixed roots were treated with 0.5 % (w/v) Cellulase Onozuka RS and 0.05 % (w/v) Pectolyase Y-23 in PIPES buffer for 45 s at 37 °C. Roots were stained with Hoechst solution (5 µg/mL Hoechst 33342 in PIPES buffer) for 10 min. Samples were visualized using a DeltaVision microscope with Olympus IX70 stand (Personal DV; Applied Precision). Images were deconvoluted using the constrained iterative algorithm implemented in SoftWoRx software (Applied Precision). Scale bar: 5 µm.

Contributors

Yuki Sakamoto*, Shingo Takagi, Department of Biological Science, Graduate School of Science, Osaka University, Machikaneyama-cho 1-1, Toyonaka, Osaka 560-0043, Japan
*E-mail: sakamoto@bio.sci.osaka-u.ac.jp

References

4. Dittmer TA, Stacey NJ, Sugimoto-Shirasu K, Richards EJ (2007) LITTLE NUCLEI genes affecting nuclear morphology in *Arabidopsis thaliana*. Plant Cell 19:2793–2803. doi:10.1105/tpc.107.053231
5. Sakamoto Y, Takagi S (2013) LITTLE NUCLEI 1 and 4 regulate nuclear morphology in *Arabidopsis thaliana*. Plant Cell Physiol 54:622–633. doi:10.1093/pcp/pct031

1 Nuclei and Chromosomes

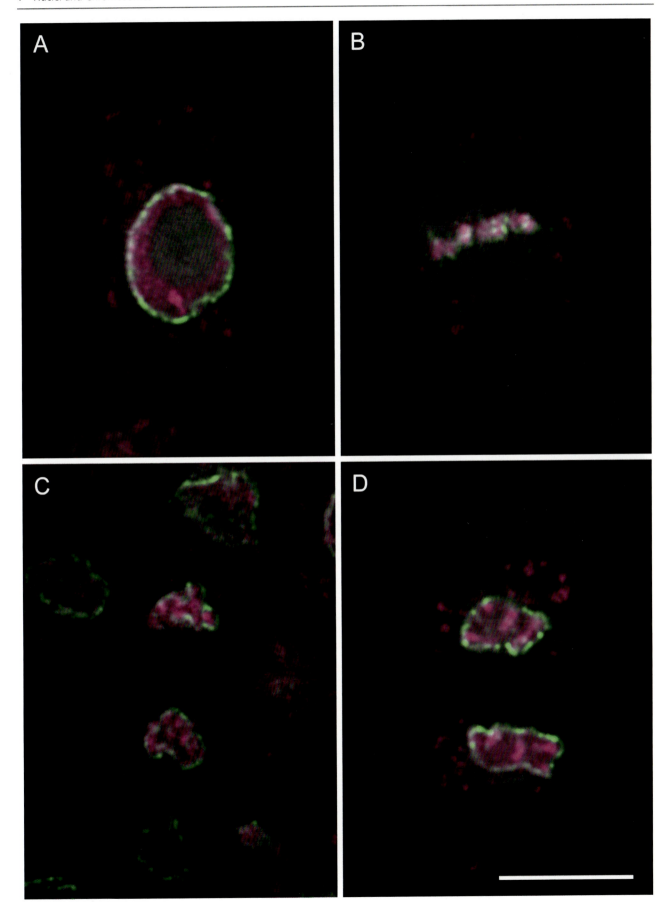

Plate 1.4

Nuclei of multinucleate cells in *Hydrodictyon reticulatum*

Hydrodictyon reticulatum (L.) Lagerheim is a free-floating freshwater green alga that forms net-like colonies with a polygonal or hexagonal mesh pattern. During net formation, hundreds of zoospores adhere to one another to form a beautiful network within the parental cell [6]. Recently adhered net-cells, which originated from zoospores, contain one nucleus (blue) (**A**). Each net-cell continues to grow larger, repeating nuclear division without cytokinesis. Large multinucleate net-cells display regularly spaced nuclei (blue) in a stationary cytoplasm (**B–D**).

Mitosis in net-cells of *H. reticulatum* is intranuclear. The nuclear envelope remains essentially intact throughout mitosis with polar fenestrae. In smaller net-cells, nuclei divide once in a 24 h period, and all nuclei are in the same phase of mitosis at the same time. In larger net-cells, mitosis occurs synchronously in waves from one end of the cell to the other (**B**). Thus, several stages of mitosis can be observed in a single cell at one time (**C, D**).

The arrangement of microtubules, chromatin and chromosomes during mitosis was examined by immunofluorescence microscopy (**C, D**) [7]. When mitosis begins, microtubules (red) radiate from each nucleus (blue) in prophase, then converge on opposed perinuclear sites and form mitotic spindles in metaphase. Spindles start to elongate in anaphase, and interzonal spindles thin and elongate as nuclei move farther apart in telophase.

Cells were treated with S-buffer containing 0.25 % glutaraldehyde and 1 μg/mL 4′, 6-diamidino-2-phenylindole (DAPI), and examined under a fluorescence microscope with an ultraviolet excitation filter (**A, B**). For microtubule immunolocalization, cells were rapidly frozen in liquid propane and transferred to chilled methanol at −80 °C. After 24 h at −80 °C, they were gradually warmed to room temperature, and incubated in blocking solution for 1 h. Samples were incubated with a monoclonal anti-α-tubulin antibody diluted 1:500 in PBS containing 1 % BSA at room temperature, then washed with PBS. Alexa568-conjugated goat anti-mouse IgG was used as the secondary antibody. After washing with PBS, samples were mounted with a mounting solution containing 1 μg/mL DAPI and examined with a confocal laser scanning microscope (**C, D**). Scale bars: 5 μm (**A**), 10 μm (**B–D**)

Contributors

Manabu Tanaka, Kyoko Hatano*, Department of Interdisciplinary Environment, Graduate School of Human and Environmental Studies, Kyoto University, Yoshida-nihonmatsu-cho, Sakyo-ku, Kyoto 606-8501, Japan
*E-mail: hatano.kyoko.7m@kyoto-u.ac.jp

References

6. Hatano K, Maruyama K (1995) Growth pattern of isolated zoospores in *Hydrodictyon reticulatum* (Chlorococcales, Chlorophyceae). Phycol Res 43:105–110
7. Tanaka M, Hatano K (2009) γ-Tubulin and microtubule organization during development of multinuclear cells and formation of zoospores in *Hydrodictyon reticulatum*. Phycologia 48(Suppl):128–129

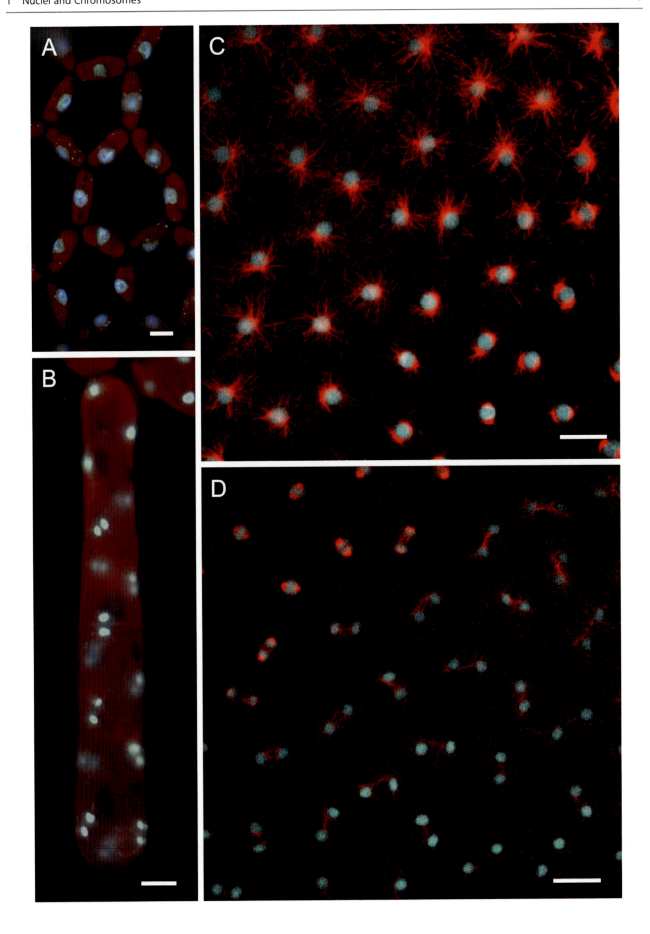

Plate 1.5

Meiotic chromosomes of *Arabidopsis thaliana* pollen mother cells

Sexually reproducing *A. thaliana* generates haploid gametes through meiosis. Meiosis is different from mitosis in that it not only do the resulting cells have half the diploid chromosome number, but also in that homologous chromosomes recombine. The meiosis-specific behavior of chromosomes is observed during prophase I and metaphase I. After one round of DNA replication in the premeiotic S phase, pollen mother cells go through G2 phase and enter prophase I. When stained with 4',6-diamidino-2-phenylindole (DAPI), leptotene *A.thaliana* chromosomes appear as entangled thin thread-like structures (**A**). Through zygotene stages, homologous chromosomes pair and synaptonemal complexes are constructed between them, elongating a thick thread-like synapsed region and leaving a thin unpaired region (**B**). It is during this stage that recombination initiated by double-stranded DNA breaks occurs. The construction of synaptonemal complexes is complete at the pachytene stage. Crossovers become visible on diplotene chromosomes, but at diakinesis chromosomes are so condensed that chromosome structures are no longer distinguishable (**C**). At metaphase I, the five bivalents congress on the metaphase plate, as opposed to the assembly of ten univalents at mitosis (**D**). Coincidental segregation of homologous chromosomes to opposite poles represents the start of anaphase I. These fluorescent images were adapted from [8]. To accomplish meiosis-specific tasks, PMCs require meiosis-specific genes.

Cells for this figure were prepared using the DAPI staining method [8], which is simple and can be easily used in further experiments, such as fluorescent in situ hybridization. Inflorescences were fixed with Farmer's solution for 24 h at room temperature, and digested with 0.3 % cytohelicase, cellulase, pectolyase in 10 mM citrate buffer (pH 4.5) for 3 h at 37 °C. A bud was transferred into a drop of 60 % acetic acid on a slide glass and incubated for 1 m. Chromosomes were the spread on the surface of the slide glass and air dried after washing with Farmer's solution. Chromosomes were stained with 1.5 µg/mL DAPI and observed under fluorescence microscope. Scale bars: 10 µm.

References

8. Azumi Y, Suzuki H (2004) New findings on molecular mechanisms of plant meiosis obtained through analyses of tagged mutants. Plant Morphol 16:31–60. doi:10.5685/plmorphol.16.31

Contributors

Yoshitaka Azumi*, Department of Biological Sciences, Faculty of Science, Kanagawa University, 2946 Tsuchiya, Hiratsuka, Kanagawa 259-1293, Japan
*E-mail: adumiy01@kanagawa-u.ac.jp

1 Nuclei and Chromosomes

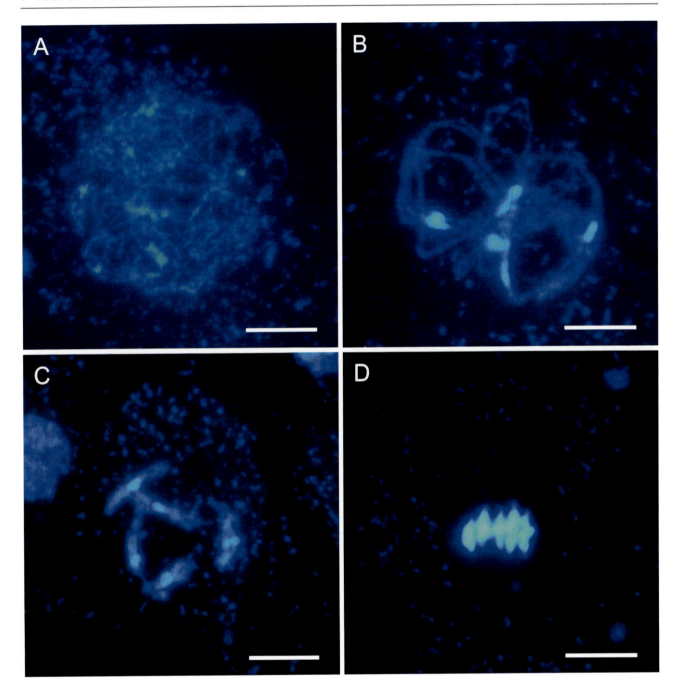

Plate 1.6

Multicolor FISH of *Pinus* chromosomes

The genus *Pinus* is a major components of Northern Hemisphere timber forests. The karyotype analyses of *Pinus* species confirm that chromosomes comprising a karyotype exhibit similar morphology. The difficulty in identification of homologous pairs prevents precise comparative karyotype analyses of *Pinus* species. Fluorescence in situ hybridization (FISH) has contributed to the karyotype analysis of *Pinus* species. FISH can identify the homologous chromosome pairs in *Pinus*.

The image adapted from [9] shows FISH of somatic chromosomes of *P. densiflora*, with each chromosome showing unique probe patterns. Color code: magenta, PCSR; green, telomere sequence; red, 18S and 5S rDNA. The *Arabidopsis*-type telomere sequence repeats $(TTTAGGG)_n$ were amplified by PCR with $(TTTAGGG)_5$ and $(CCCTAAA)_5$ primers without template DNA and labeled with biotin. 18S rDNA and 5S rDNA probes were labeled with digoxigenin (DIG). Plasmid DNA containing *PCSR* (Proximal CMA band-Specific Repeat, clone PDCD501; accession no. AB051860) [10] was labeled with FITC. Probes were dissolved in a solution of $2\times$ SSC, 10 % dextran sulfate, and 50 % formamide. Chromosomal DNA was denatured at 80 °C for 1 min in 70 % formamide, $2\times$ SSC. Hybridized probes were detected with Strepetavidin-Cy5 and anti-digoxigenin antibody conjugated with rhodamine and slides were counterstained with DAPI (4, 6-diamino-2-phenylidole). Hybridization signals were visualized and recorded using a chilled CCD camera (Sensys 1400, Photometrics), and pseudocolor images were made using IPLab (Scanalytics). Scale bars: 10 µm.

Contributors

Masahiro Hizume*, Fukashi Shibata, Faculty of Education, Ehime University, 3 Bunkyo, Matsuyama 790-8577, Japan
*E-mail: hizume@ed.ehime-u.ac.jp

References

9. Hizume M, Shibata F, Matsusaki Y, Garajova Z (2002) Chromosome identification and comparative karyotypic analyses of four *Pinus* species. Theor Appl Genet 105:491–497. doi:10.1007/s00122-002-0975-4
10. Hizume M, Shibata F, Maruyama Y, Kondo T (2001) Cloning of DNA sequences localized on proximal fluorescent chromosome bands by microdissection in *Pinus densiflora* Sieb. & Zucc. Chromosoma 110:345–351. doi:10.1007/s004120100149

1 Nuclei and Chromosomes

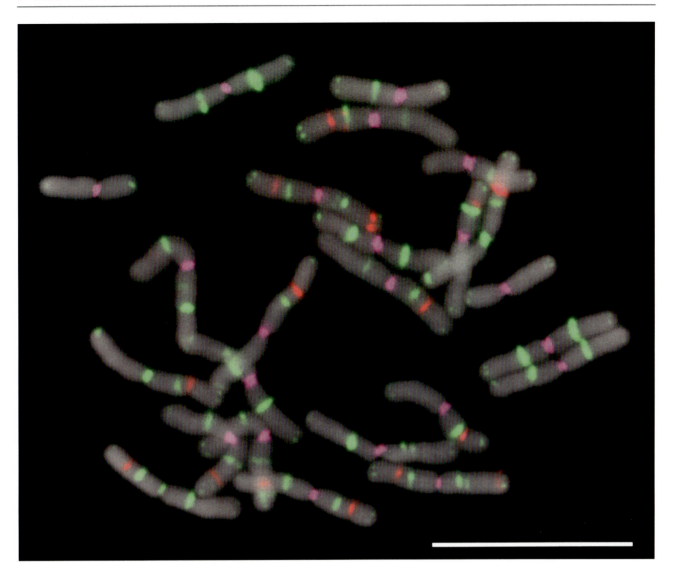

Plate 1.7

Chromosome painting and FISH of distal end satellite DNAs in dioecious plants with sex chromosomes

The dioecious plant *Rumex acetosa* has a multiple sex chromosome system: females are $2n = XX + 12$, males are $2n = XY_1Y_2 + 12$. Two DNA sequences which are abundant on Y chromosomes have been isolated, one is the Y chromosome specific sequence (RAYSI) [11] and the other is located on Y chromosomes and a pair of autosomes (RAE180) [12]. These two repetitive sequences are present in unique, complex patterns on Y_1 and Y_2 [12]. Two bright signals corresponding to these Y abundant repetitive sequences are detected on two Y chromosomes of male *R. acetosa* prometapahase chromosomes probed by fluorescence in situ hybridization (**A**). The image was adapted from [11].

Silene latifolia is also a dioecious plant with heteromorphic X and Y sex chromosomes. The recombining pseudoautosomal regions (PAR) are required for proper sex chromosome division in meiosis. PAR localization on the Y can be detected by FISH with a probe against the *Sac*I satellite subfamily [13]. FISH shows 25 chromosomes in early metaphase of a male *S. latifolia* aneuploidy line (**B**). FISH signals are detected on most of autosomes, both arms of the X, and the non-condensed arm of the Y [14]. Since homologous recombination occurs in PARs, PAR DNA sequences should be similar on both X and Y chromosomes, indicating that the PAR is located on the non-condensed arm of the Y chromosome. The image was adapted from [14].

DOP-PCR amplified Y-DNA of *R. acetosa* was labeled with biotin by nick translation. Labeled DNA was resuspended in hybridization solution (50 % formamide, 10 % dextran sulfate, 0.08 mol/L Na_2HPO_4 in $2\times$ SSC pH 6.5) with unlabeled genome DNA. Chromosomes were denatured in 70 % formamide/$2\times$ SSC at 76 °C for 60 s. Hybridized probes were detected using avidin-FITC (EY Laboratories). Slides were counterstained with propidium iodide.

*Sac*I satellite subfamily probe was prepared by direct labeling with Alexa Fluor 546. Chromosomal DNA was denatured at 75 °C for 1 min in 70 % formamide/$2\times$ SSC. Chromosome preparations were dehydrated immediately using 5 min treatment with 70 % ethanol at -20 °C followed by 100 % ethanol at room temperature. Preparations were then treated with acetone for 30 min at room temperature. Slides were washed at 42 °C in 50 % formamide/$2\times$ SSC and conterstained with DAPI. Scale bars: 5 μm.

Contributors

Fukashi Shibata[1], Yusuke Kazama[2]*, Shigeyuki Kawano[3], Masahiro Hizume[1†], [1]Faculty of Education, Ehime University, 3 Bunkyo, Matsuyama 790-8577, Japan, [2]Ion Beam Breeding Team, RIKEN Nishina Center, 2-1 Hirosawa, Wako, Saitama 351-0198, Japan, [3]Department of Integrated Biosciences, Graduate School of Frontier Sciences, The University of Tokyo, Bldg. FSB-601, 5-1-5 Kashiwanoha, Kashiwa, Chiba 277-8562, Japan
*E-mail: ykaze@riken.jp, †hizume@ed.ehime-u.ac.jp

References

11. Shibata F, Hizume M, Kuroki Y (1999) Chromosome painting of Y chromosomes and isolation of a Y chromosome-specific repetitive sequence in the dioecious plant *Rumex acetosa*. Chromosoma 108:266–270
12. Shibata F, Hizume M, Kuroki Y (2000) Differentiation and the polymorphic nature of the Y chromosomes revealed by repetitive sequences in the dioecious plant, *Rumex acetosa*. Chromosome Res 8:229–236
13. Kazama Y, Sugiyama R, Suto Y, Uchida W, Kawano S (2006) The clustering of four subfamilies of satellite DNA at individual chromosome ends in *Silene latifolia*. Genome 49:520–530
14. Kazama Y, Kawano S (2008) Technical note: detection of pseudo autosomal region in the *Silene latifoia* Y chromosome by FISH analysis of distal end satellite DNAs. Cytologia 73(ii):3–4

1 Nuclei and Chromosomes   15

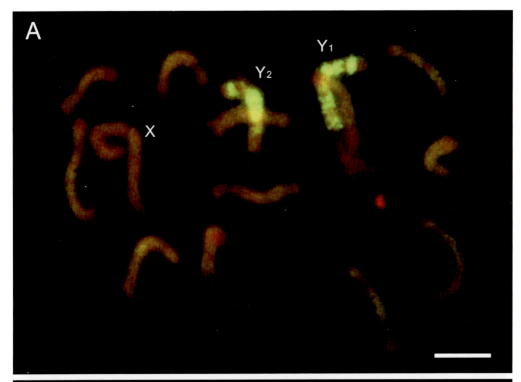

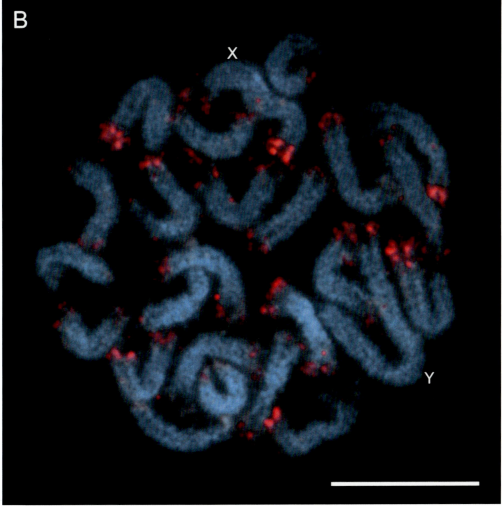

Plate 1.8

Kinetochore and microtubule dynamics during cell division of tobacco BY-2 cells visualized by live-cell imaging

Mitosis and cytokinesis are crucial for the equal separation of genetic information to both daughter cells [2]. A series of highly coordinated events such as bipolar spindle formation, and chromosome alignment, and chromosome segregation occur in mitosis. The accuracy of cytokinesis in plant cells is also ensured by phragmoplast formation.

Kinetochore and microtubule dynamics were visualized in living BY-2 cells expressing green fluorescent protein (GFP)-fused α-tubulin and tdTomato (a tandem dimer variant of red fluorescent protein [RFP])-fused CenH3 (centromeric histone H3: *Arabidopsis thaliana* HTR12) [15]. The centromeric region contains histone H3 variant CenH3. The kinetochore is established by the attachment of spindle microtubules to the centromere. Before nuclear envelope breakdown at prophase (Pro), cortical microtubules are replaced by the densely packed preprophase band (PPB). Microtubules begin organizing at prometaphase (Prometa) and kinetochores move to the spindle equator. As cells reach metaphase (Meta), CenH3s oscillate and align on the spindle equator. After complete CenH3 alignment, CenH3s segregate equally to the opposing sides. The phragmoplast emerges at the midzone at late anaphase. Lateral expansion of the phragmoplast is observed during telophase (Telo). The cell wall is formed after disappearance of the phragmoplast. After cytokinesis (Cyt) cell walls are synthesized and new cell nuclei are visible following nuclear envelope reconstruction in the two daughter cells.

Tobacco BY-2 cells on coverslips were transferred to petri dishes (Matsunami Glass Ind., Ltd.). Dishes were placed on the inverted platform of a fluorescence microscope (IX-81; Olympus) equipped with an electron multiplying cooled charged-coupled device (EMCCD) camera (Evolve 512; Photometrics), a Piezo focus drive (P-721; Physik Instrumente), and 488 nm and 561 nm lasers (Sapphire; Coherent). Images were acquired every 1 min with a 60× objective lens (UPLSAPO 60XS, Silicone Oil; Olympus). Green and magenta fluorescence represent α-tubulin and CenH3, respectively. Scale bar: 10 μm.

Contributors

Daisuke Kurihara[1]*, Sachihiro Matsunaga[2], [1]Division of Biological Science, Graduate School of Science, Nagoya University, Furo-cho, Chikusa-ku, Nagoya, Aichi 464-8602, Japan, [2]Department of Applied Biological Science, Faculty of Science and Technology, Tokyo University of Science, 2641 Yamazaki, Noda, Chiba 278-8510, Japan
*E-mail: kuri@bio.nagoya-u.ac.jp

References

2. Matsunaga S, Katagiri Y, Nagashima Y, Sugiyama T, Hasegawa J, Hayashi K, Sakamoto T (2013) New insights into the dynamics of plant cell nuclei and chromosomes. Int Rev Cell Mol Biol 305:253–301. doi:10.1016/B978-0-12-407695-2.00006-8
15. Kurihara D, Matsunaga S, Uchiyama S, Fukui K (2008) Live cell imaging reveals plant Aurora kinase has dual roles during mitosis. Plant Cell Physiol 49:1256–1261. doi:10.1093/pcp/pcn098

1 Nuclei and Chromosomes

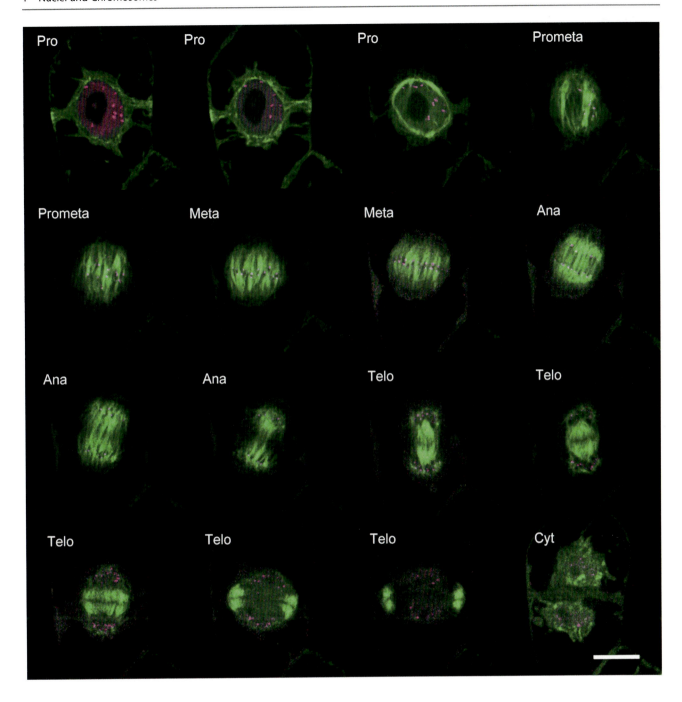

Plate 1.9

Visualization of chromatin dynamics in the root of *Arabidopsis thaliana*

Chromatin and nuclear dynamics are closely related to many biological processes in eukaryotes. Chromatin movement is associated with subnuclear events such as DNA replication, transcription, and repair. The chromatin fluorescence tagging system is a technique using the bacterial operator/repressor system combined with a fluorescence protein, for example, *lacO*/LacI-EGFP. This system can visualize specific loci where the operator tandem array is inserted [16]. FISH and immunofluorescence experiments require the fixation of cells but the system can analyze mitotic and interphase chromatin dynamics in living cells.

The left image represents chromatin dynamics in a root of *Arabidopsis thaliana* expressing H2B-tdTomato, a histone H2B protein, fused with fluorescent protein, tdTomato. The two green dots in a nucleus are *lacO*/LacI-EGFP signals which show specific loci on two homologous chromosomes. An epidermal cell at metaphase is located in the upper left region of the image. Mitotic chromosomes align at the metaphase plate, as shown by the alignment of *lacO*/LacI-EGFP signals.

A. thaliana roots consist of two regions: a meristematic region of mitotic cells and an elongation region of rapidly expanding cells [2]. In the elongation region, repeated DNA replication without cell division, endoreduplication, coupled with rapid cell expansion [17]. Nuclei are larger in the elongation region than in the meristematic region. The *lacO*/LacI-EGFP signals also enlarge in the elongation region. The distance between the two dots, the inter-allelic distance, is longer in the elongation region than in the meristematic region.

Seeds of *A. thaliana* expressing *lacO*/LacI-EGFP and H2B-tdTomato were germinated on 1/2MS medium (0.1 % sucrose, 0.6 % gellan gum) in a glass bottomed dish. The dish was placed at 4 °C for 1 day, then moved to incubator and grown at 22 °C in 16 h light/8 h night cycle. Seedling roots were observed 5 day after germination under an inverted fluorescent microscope (IX-81, Olympus) equipped with a confocal scanning unit (CSUX-1, Yokogawa) and a sCMOS camera (Neo 5.5 sCMOS ANDOR). One stack consists of 40 images at 0.5 μm z-axis steps collected for 90 min (time interval: 5 min). Images were analyzed with ImageJ software. Scale bar: 50 μm.

Contributors

Takeshi Hirakawa, Sachihiro Matsunaga*, Department of Applied Biological Science, Faculty of Science and Technology, Tokyo University of Science, 2641 Yamazaki, Noda, Chiba 278-8510, Japan
*E-mail: sachi@rs.tus.ac.jp

References

16. Matzke AJ, Watanabe K, van der Winden J, Naumann U, Matzke M (2010) High frequency, cell type-specific visualization of fluorescent-tagged genomic sites in interphase and mitotic cells of living *Arabidopsis* plants. Plant Methods 6:2. doi:10.1186/1746-4811-6-2

2. Matsunaga S, Katagiri Y, Nagashima Y, Sugiyama T, Hasegawa J, Hayashi K, Sakamoto T (2013) New insights into the dynamics of plant cell nuclei and chromosomes. Int Rev Cell Mol Biol 305:253–301. doi:10.1016/B978-0-12-407695-2.00006-8

17. Hayashi K, Hasegawa J, Matsunaga S (2013) The boundary of the meristematic and elongation zones in roots: endoreduplication precedes rapid cell expansion. Sci Rep 3:2723. doi:10.1038/srep02723

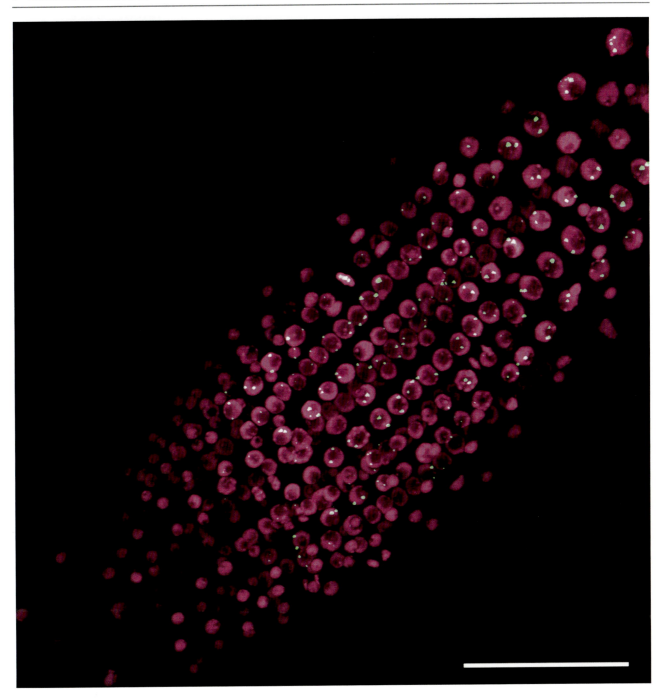

Plate 1.10

Specific contribution of condensin II to sister centromere resolution in *Cyanidioschyzon merolae*

Condensins are multisubunit complexes that play central roles in chromosome organization and segregation in eukaryotes. Many eukaryotic species, including humans and flowering plants, have two different condensin complexes (condensins I and II), which leads to many interesting questions. Why do the two types of condensins exist? Do they have unique functions? To address these questions, the red alga *Cyanidioschyzon merolae* was used because it represents the smallest and simplest organism that is known to possess condensins I and II. Despite the great evolutionary distance, spatiotemporal dynamics of condensins in *C. merolae* and mammalian cells are strikingly similar: condensin II localizes to the nucleus throughout the cell cycle, whereas condensin I becomes visible on chromosomes only after the nuclear envelope partially dissolves at prometaphase. Unlike in mammalian cells, however, condensin II is concentrated at centromeres in metaphase, whereas condensin I is distributed more broadly along chromosome arms. A targeted gene disruption technique was established in *C. merolae* [18], and this technique demonstrated that condensin II is not essential for mitosis under laboratory growth conditions. Condensin II does, however, play a crucial role in facilitating sister centromere resolution in the presence of a microtubule drug. These results offer an excellent example of how the combination of genetics and cell biology in *C. merolae* can provide fundamental insights into basic cellular processes such as chromosome architecture and dynamics [19].

Shown here is a phase-contrast image of *C. merolae* metaphase cells immunolabeled with an antibody against centromere-specific histone H3 (CenH3) (light blue) adapted from [19]. The contour of a wild-type cell is pseudo-colored blue; the contours of condensin II–knockout cells are pseudo-colored red. In the wild-type cell, two CenH3 clusters are well resolved at metaphase in the presence or absence of functional microtubules. In condensin II–knockout cells, however, sister centromere clusters fail to resolve in the absence of functional microtubules. Scale bar: 1 μm. This image was chosen as the cover of *Molecular Biology of the Cell* (vol 24, no. 16).

Contributors

Takayuki Fujiwara, Tatsuya Hirano*, Chromosome Dynamics Laboratory, RIKEN, Wako, Saitama 351-0198, Japan
*E-mail: hiranot@riken.jp

References

18. Fujiwara T, Ohnuma M, Yoshida M, Kuroiwa T, Hirano T (2013) Gene targeting in the red alga *Cyanidioschyzon merolae*: single- and multi-copy insertion using authentic and chimeric selection marker. PLoS One 8:e73608. doi:10.1371/journal.pone.0073608

19. Fujiwara T, Tanaka K, Kuroiwa T, Hirano T (2013) Spatiotemporal dynamics of condensins I and II: evolutionary insights from the primitive red alga *Cyanidioschyzon merolae*. Mol Biol Cell 16:2515–2527. doi:10.1091/mbc.E13-04-0208

1 Nuclei and Chromosomes

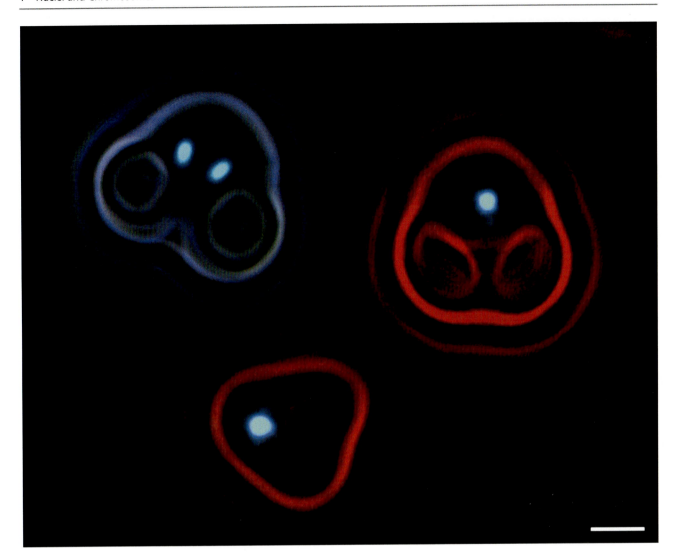

Plate 1.11

Endomitosis induces a giant polyploid cell on the leaf epidermis

An increase in ploidy can be caused by one of two events, endoreduplication or endomitosis, whereby chromosomes replicate but cells do not divide. Endoreduplication, in which cells skip the entire processes of mitosis, does not increase the number of chromosomes but generates polytene chromosomes. Cells in endomitosis enter but do not finish mitosis, proceeding through anaphase without nuclear division and cytokinesis. In contrast to endoreduplication, endomitosis causes doubling of the chromosome number [2].

GIGAS CELL1 (GIG1) encodes a plant-specific inhibitor of anaphase-promoting complex/cyclosome (APC/C) in *Arabidopsis thaliana* [20]. APC/C is responsible for the transition between mitotic processes by degradation of cell cycle proteins. GIG1 prevents the ectopic occurrence of endomitosis by inhibiting APC/C. The recessive *gig* mutat was identified as an enhancer of the *myb3r4* mutant phenotype. MYB3R4 is a member of the Myb family of transcriptional regulators that positively regulate mitotic progression in *A. thaliana*. Giant cells, called *gigas* cells, are observed in the epidermis of *gig1* cotyledons. The frequency of *gigas* cells is increased when MYB3R4 and GIG1 are simultaneously mutated.

The left figure shows a fluorescent image of the epidermis of cotyledons in *gig1/myb3r4* double mutants expressing TOO MANY MOUTHS (TMM)-GFP (green) and tdTomato-CENH3 (red). TMM is a marker for stomatal precursor cells and is essential for stomatal development. The *gigas* cell in the center of the image strongly expresses TMM, suggesting that *gigas* cells may have a guard cell–like identity that arises from a developmental pathway similar to the one that generates stomata. The nuclei of *gigas* cells are larger than those of normal guard cells, and their precursors and are equivalent in size to endoreduplicated nuclei in pavement cells (jigsaw puzzle–shaped cells). Ploidy level can be estimated by the number of chromosomes in each nucleus using a kinetochore-specific marker, tdTomato-CENH3. Guard cells show 10 tdTomato-CENH3 signals, which is equivalent to the diploid chromosome number (2n = 10) of *A. thaliana*. In contrast, the nuclei of *gigas* cells contain 20 signals. This suggests that *gigas* cells have 20 chromosomes which are generated by endomitosis.

Imaging was performed using an inverted fluorescence microscope (IX-81; Olympus) equipped with a confocal laser scanner unit CSUX-1 (Yokogawa Electronic) and a charge-coupled device camera (CoolSNAP HQ2; Roper Scientific). Scale bar: 50 μm.

Contributors

Sachihiro Matsunaga[1]*, Masaki Ito[2], [1]Department of Applied Biological Science, Faculty of Science and Technology, Tokyo University of Science, 2641 Yamazaki, Noda, Chiba 278-8510, Japan, [2]Graduate School of Bioagricultural Sciences, Nagoya University, Chikusa, Nagoya 464-8601, Japan
*E-mail: sachi@rs.tus.ac.jp

References

2. Matsunaga S, Katagiri Y, Nagashima Y, Sugiyama T, Hasegawa J, Hayashi K, Sakamoto T (2013) New insights into the dynamics of plant cell nuclei and chromosomes. Int Rev Cell Mol Biol 305:253–301. doi:10.1016/B978-0-12-407695-2.00006-8

20. Iwata E, Ikeda S, Matsunaga S, Kurata M, Yoshioka Y, Criquid M-C, Genschik P, Ito M (2011) GIGAS CELL1, a novel negative regulator of APC/C, is required for proper mitotic progression and cell fate determination in *Arabidopsis thaliana*. Plant Cell 23:4382–4393. doi:10.1105/tpc.111.092049

1 Nuclei and Chromosomes

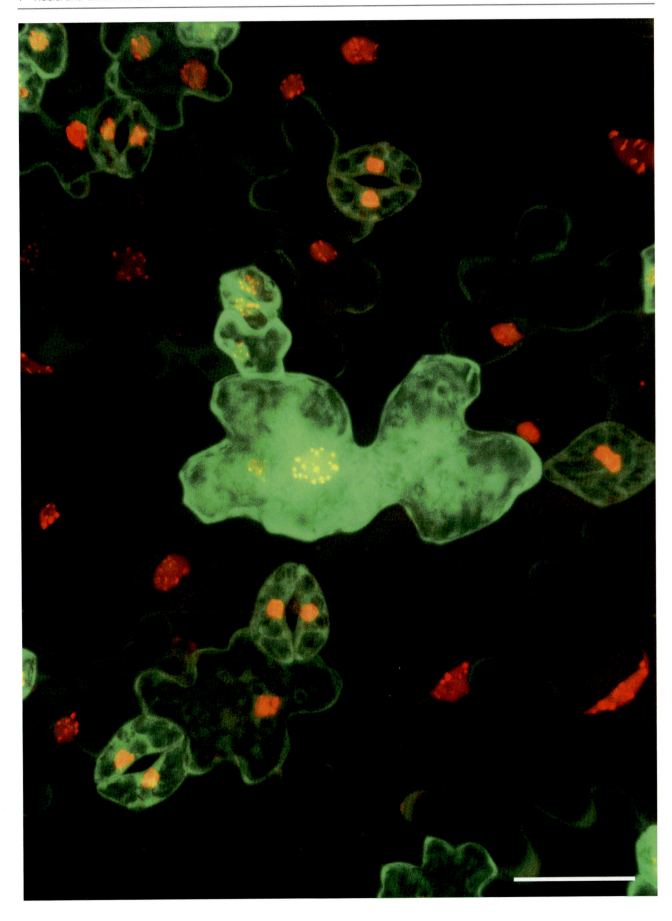

Chapter References

1. Karahara I, Suda J, Tahara H, Yokota E, Shimmen T, Misaki K, Yonemura S, Staehelin A, Mineyuki Y (2009) The preprophase band is a localized center of clathrin-mediated endocytosis in late prophase cells of the onion cotyledon epidermis. Plant J 57:819–831. doi:10.1111/j.1365-313X.2008.03725.x
2. Matsunaga S, Katagiri Y, Nagashima Y, Sugiyama T, Hasegawa J, Hayashi K, Sakamoto T (2013) New insights into the dynamics of plant cell nuclei and chromosomes. Int Rev Cell Mol Biol 305:253–301. doi:10.1016/B978-0-12-407695-2.00006-8
3. Matsunaga S, Fukui K (2010) The chromosome peripheral proteins play an active role in chromosome dynamics. Biomol Concepts 1:157–164. doi:10.1515/bmc.2010.018
4. Dittmer TA, Stacey NJ, Sugimoto-Shirasu K, Richards EJ (2007) LITTLE NUCLEI genes affecting nuclear morphology in *Arabidopsis thaliana*. Plant Cell 19:2793–2803. doi:10.1105/tpc.107.053231
5. Sakamoto Y, Takagi S (2013) LITTLE NUCLEI 1 and 4 regulate nuclear morphology in *Arabidopsis thaliana*. Plant Cell Physiol 54:622–633. doi:10.1093/pcp/pct031
6. Hatano K, Maruyama K (1995) Growth pattern of isolated zoospores in *Hydrodictyon reticulatum* (Chlorococcales, Chlorophyceae). Phycol Res 43:105–110
7. Tanaka M, Hatano K (2009) γ-Tubulin and microtubule organization during development of multinuclear cells and formation of zoospores in *Hydrodictyon reticulatum*. Phycologia 48 (Suppl):128–129
8. Azumi Y, Suzuki H (2004) New findings on molecular mechanisms of plant meiosis obtained through analyses of tagged mutants. Plant Morphol 16:31–60. doi:10.5685/plmorphol.16.31
9. Hizume M, Shibata F, Matsusaki Y, Garajova Z (2002) Chromosome identification and comparative karyotypic analyses of four *Pinus* species. Theor Appl Genet 105:491–497. doi:10.1007/s00122-002-0975-4
10. Hizume M, Shibata F, Maruyama Y, Kondo T (2001) Cloning of DNA sequences localized on proximal fluorescent chromosome bands by microdissection in *Pinus densiflora* Sieb. & Zucc. Chromosoma 110:345–351. doi:10.1007/s004120100149
11. Shibata F, Hizume M, Kuroki Y (1999) Chromosome painting of Y chromosomes and isolation of a Y chromosome-specific repetitive sequence in the dioecious plant *Rumex acetosa*. Chromosoma 108:266–270
12. Shibata F, Hizume M, Kuroki Y (2000) Differentiation and the polymorphic nature of the Y chromosomes revealed by repetitive sequences in the dioecious plant, *Rumex acetosa*. Chromosome Res 8:229–236
13. Kazama Y, Sugiyama R, Suto Y, Uchida W, Kawano S (2006) The clustering of four subfamilies of satellite DNA at individual chromosome ends in *Silene latifolia*. Genome 49:520–530
14. Kazama Y, Kawano S (2008) Technical note: detection of pseudo autosomal region in the *Silene latifoia* Y chromosome by FISH analysis of distal end satellite DNAs. Cytologia 73(ii):3–4
15. Kurihara D, Matsunaga S, Uchiyama S, Fukui K (2008) Live cell imaging reveals plant Aurora kinase has dual roles during mitosis. Plant Cell Physiol 49:1256–1261. doi:10.1093/pcp/pcn098
16. Matzke AJ, Watanabe K, van der Winden J, Naumann U, Matzke M (2010) High frequency, cell type-specific visualization of fluorescent-tagged genomic sites in interphase and mitotic cells of living *Arabidopsis* plants. Plant Methods 6:2. doi:10.1186/1746-4811-6-2
17. Hayashi K, Hasegawa J, Matsunaga S (2013) The boundary of the meristematic and elongation zones in roots: endoreduplication precedes rapid cell expansion. Sci Rep 3:2723. doi:10.1038/srep02723
18. Fujiwara T, Ohnuma M, Yoshida M, Kuroiwa T, Hirano T (2013) Gene targeting in the red alga *Cyanidioschyzon merolae*: single- and multi-copy insertion using authentic and chimeric selection marker. PLoS One 8:e73608. doi:10.1371/journal.pone.0073608
19. Fujiwara T, Tanaka K, Kuroiwa T, Hirano T (2013) Spatiotemporal dynamics of condensins I and II: evolutionary insights from the primitive red alga *Cyanidioschyzon merolae*. Mol Biol Cell 16:2515–2527. doi:10.1091/mbc.E13-04-0208
20. Iwata E, Ikeda S, Matsunaga S, Kurata M, Yoshioka Y, Criquid M-C, Genschik P, Ito M (2011) GIGAS CELL1, a novel negative regulator of APC/C, is required for proper mitotic progression and cell fate determination in *Arabidopsis thaliana*. Plant Cell 23:4382–4393. doi:10.1105/tpc.111.092049

Mitochondria

2

Shigeyuki Kawano

Mitochondria are responsible for respiration and energy generation in eukaryotes. The double-membraned organelle contains cristae and its own DNA (mtDNA), and is inherited from parent to progeny by fission and appropriate distribution. The packaging of mtDNA into DNA-protein assemblies called nucleoids provides an efficient segregating unit of mtDNA, coordinating mtDNA's involvement in cellular metabolism and distribution to daughter mitochondria upon fission. The mitochondrial nucleoid is composed of a set of DNA-binding core proteins involved in mtDNA maintenance and transcription. Although mtDNA is inherited from parent to progeny much like nuclear DNA through sexual reproduction, a wide range of species exhibit restricted inheritance of mtDNA from only one parent to one progeny.

In this chapter, T. Kuroiwa and I. Miyakawa present remarkable studies of mitochondrial dynamics based on visualization using mtDNA and mitochondrial nucleoids resulting from collaborations with experts in various fields of biology. Many interesting discoveries have been accomplished in organellar division and inheritance. N. Sasaki visualizes mtDNA and mitochondrial nucleoids in human cells and the slime mold *Physarum polycephalum* using fluorescent dye. Y. Moriyama and S. Kawano present the behavior of mtDNA and the mitochondrial nucleoid during mating of a different slime mold, *Didymium iridis*, and demonstrate digestion of mtDNA in the process of maternal inheritance of mitochondria.

The shape and size of mitochondria are highly variable between organisms, and their dynamics are driven by movement, fission, and fusion. Transmission electron microscopy (TEM) has enabled major progress in the characterization and understanding of the morphology and ultrastructure of mitochondria, including the presence of double membranes and cristae. An advanced TEM technology is the reconstruction of a 3D image from a series of TEM images, which provides many insights into the structural differences of mitochondria between different cell types. On the other hand, fluorescence microscopy is an excellent method for analyzing mitochondrial dynamics. Especially important is the discovery that certain lipophilic fluorescent dyes stain mitochondria in many types of living cells. Not only is the staining of mtDNA with DNA-specific fluorescent dyes a good method to visualize mitochondria in cells, but this technique also allows scientists to monitor mtDNA dynamics during several biological processes. The use of green fluorescent protein (GFP) to visualize a particular protein in living cells has been applied to research on mitochondrial dynamics. Some GFP variants have unique optical properties which can be used to powerful effect by cell biologists.

In this chapter, T. Ehara and T. Osafune describe mitochondrial structure during the cell cycle of *Chlamydomonas* using 3D TEM reconstruction technology. Dynamics of mitochondria and mtDNA during germination in *Chlamydomonas* are shown by H. Aoyama et al. Interestingly, *Euglena*, a genus of unicellular flagellate protists, have a cellular network of mitochondria containing approximately 150 nucleoids, according to Y. Hayashi and K. Ueda. S. Arimura revealed the dynamics of more than 10,000 mitochondria in onion epidermal cells by using a photoconvertible fluorescent GFP variant protein Kaede.

Why do plants need both chloroplasts and mitochondria? Plants perform two vital energy conversion and storage functions. Photosynthesis is the conversion of solar energy into chemical energy in the form of glucose, and takes place in chloroplasts. At the same time, plants need energy for growth and other metabolic processes. This energy is produced by mitochondria using glucose and oxygen. Although plant mitochondria as in other eukaryotes play an essential role in the cell as the major producers of ATP by oxidative phosphorylation, they also play crucial roles in many other aspects of plant development, and possess some unique properties which allow them function in plant cell metabolism. In this chapter, C. Saito shows that *Arabidopsis* contains spherical mitochondrion in guard cells, as visualized by fluorescence imaging. Skunk cabbage generates heat in the inflorescence during reproductive organ development. M. Sato et al. demonstrate by TEM that this plant has a large number of mitochondria in its thermogenesic organ, suggesting the possibility that mitochondria may contribute to thermogenesis.

S. Kawano (✉)
Department of Integrated Biosciences, Graduate School of Frontier Sciences, The University of Tokyo, 5-1-5 Kashiwanoha, Kashiwa, Chiba 277-8562, Japan
e-mail: kawano@k.u-tokyo.ac.jp

T. Noguchi et al. (eds.), *Atlas of Plant Cell Structure*,
DOI 10.1007/978-4-431-54941-3_2, © Springer Japan 2014

Plate 2.1

Mechanisms of division and inheritance of mitochondria and chloroplasts

To elucidate the universality of the mechanism of mitochondrial (mt) division previously found in *Physarum polycephalum*, a high resolution fluorescence microscopy method was developed [1]. Using this method, it was possible to observe a ribosomal gene in the tobacco chloroplast genome, T4 phage, and image mt-nuclear division. The examination of the dynamics of cell nuclei and mtDNA in *Saccharomyces cerevisiae* revealed that many small mitochondria and their nuclei fuse to form a mesh-like mitochondrion network containing mt-nuclei during mitosis and meiosis [1]. Images from these studies were chosen as both the cover of *Nature* (**A**) [2] and the cover of the book published by CSHL PRESS (**B**) [3]. The latter was the first identification of 16 chromosomes and their DNA contents in each S. cerevisiae chromosome (**C**) [4, 5]. These events were also cited in a molecular biology textbook [6]. While studying the dynamics of cell nuclei and chloroplast (Ch) nuclei (nucleoids) in *Chlamydomonas reinhardtii* and *Acetabularia calyculus*, it was discovered that cp-nuclei from the male parent were selectively digested in 1 h after mating (**D**), suggesting the mechanism of maternal inheritance of cp-genes. This image of *C. reinhardtii* zygotes was chosen as the cover of *Nature* [7] and the finding has been characterized as a universal mechanism of uniparental inheritance of organelles in eukaryotes [8]. Scale bars: 2 µm. This figure is adapted from [2–4, 7].

Contributors

Tsuneyoshi Kuroiwa[1]*, Isamu Miyakawa[2], [1]CREST, Initiative Research Unit, College of Science, Rikkyo University, Toshima, Tokyo 171-8501, Japan, [2]Faculty of Science, Yamaguchi University, Yoshida 1677-1, Yamaguchi 753-8512, Japan
*E-mail: tsune.kuroiwa@gmail.com

References

1. Kuroiwa T (1982) Mitochondrial nuclei. Int Rev Cytol 75:1–59
2. Miyakawa N, Aoi H, Sando N, Kuroiwa T (1991) Mitochondrial import proteins. Nature 349(6306)
3. Kuroiwa T, Miyakawa I (1991) In: Broach JR et al (eds) The molecular and cellular biology of the yeast *Saccharomyces*. Cold Spring Harbor Laboratory Press, New York
4. Kuroiwa T, Kojima H, Miyakawa, I, Sando N (1984) Meiotic karyotype of the yeast *Saccharomyces cerevisiae*. Exp Cell Res 153:259–265
5. Kuroiwa T, Miyamura S, Kawano S, Hizume M, Toh-e A, Miyakawa I, Sando N (1986) Cytological characterization of a synaptonemal complex-less nucleolar-organizing region on a bivalent in the bivalent of *Saccharomyces cerevisiae* using a video-intensified microscope system. Exp Cell Res 165:199–206
6. Watson JD, Hopkins NH, Roberts JW, Steitz JA, Weiner AM (1987) Molecular biology of the gene, 4 edn. Benjamin Cummings, San Francisco, pp 552–553
7. Kuroiwa T, Kawano S, Nishibayashi S, Sato C (1982) Maternal inheritance of chloroplast DNA. Nature 298(5873):481–483
8. Kuroiwa T (2010) Review of cytological studies on cellular and molecular mechanisms of uniparental (maternal or paternal) inheritance of plastid and mitochondrial genomes induced by active digestion of organelle nuclei (nucleoids). J Plant Res 123:207–230

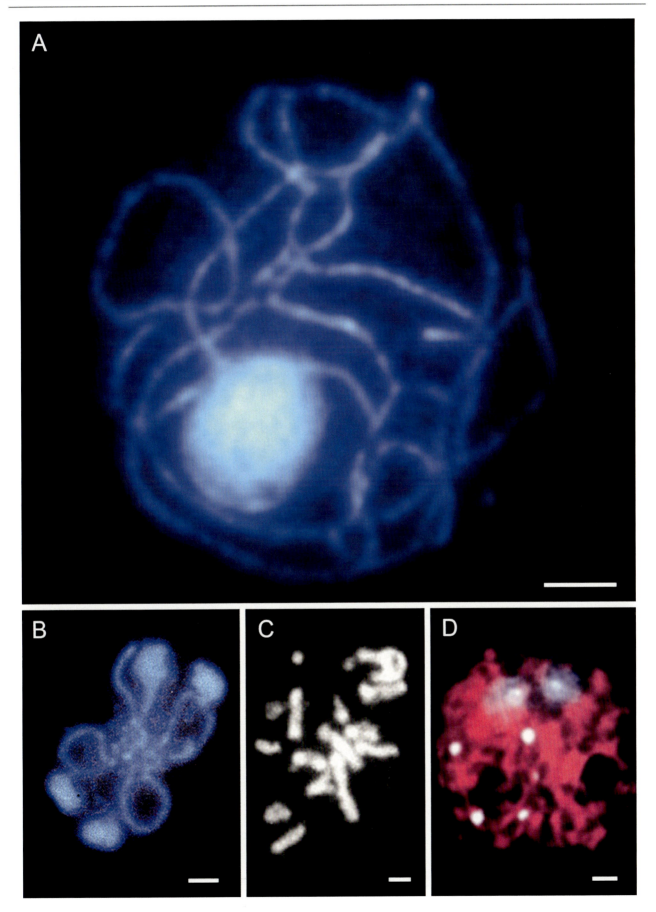

Plate 2.2

Mitochondrial nucleoid of *Physarum polycephalum*

Mitochondria have their own DNA (mtDNA). Under electron microscopy, a small amount of DNA-like fibers appear in an electron transparent or semi-transparent spherical mitochondrial area. As a result of electron microscopic observations, it was believed that mtDNA, unlike nuclear DNA, was not packaged with proteins. In the 1970s, Kuroiwa showed that mtDNA is organized in a compact structure with proteins, called mitochondrial nucleoid (mt-nucleoid) or nuclei (mt-nuclei) in the slime mold *Physarum polycephalum* [1]. *P. polycephalum* has extraordinary large rod-shaped mt-nucleoids which are easily observed as electron-dense structures at the center of mitochondria under electron microscopy. Each mt-nucleoid of *P. polycephalum* contains 40–80 copies of 63-kbp mtDNA. Comparing to *P. polycephalum*, the amount of mtDNA in mt-nucleoids of other organisms is much smaller. For example, each human mt-nucleoid contains about 3–4 copies of 16.6-kbp mtDNA. This figure shows mt-nucleoids in a human cell (**left**) and *P. polycephalum* (**right**).

P. polycephalum amoeba were fixed with 0.6 % glutaraldehyde and 0.04 % Triton X-100 and stained with 4′,6-diamidino-2-phenylindole (DAPI). HeLa cells grown on a coverslip were incubated in culture medium containing SYBR Green I (diluted 1:100,000) at 37 °C in humidified 5 % CO_2 and washed twice with culture medium [9]. After staining, cells were observed under ultraviolet and blue excitation with an epifluorescence microscope, for DAPI- and SYBR Green I-staining respectively. Scale bar: 5 μm.

Contributors

Narie Sasaki*, Division of Biological Science, Graduate School of Science, Nagoya University, Nagoya, Aichi 464-8602, Japan
*E-mail: narie@bio.nagoya-u.ac.jp

References

1. Kuroiwa T (1982) Mitochondrial nuclei. Int Rev Cytol 75:1–59
9. Ozawa S, Sasaki N (2009) Visualization of nucleoids in living human cells using SYBR Green I. Cytologia 74:365–366

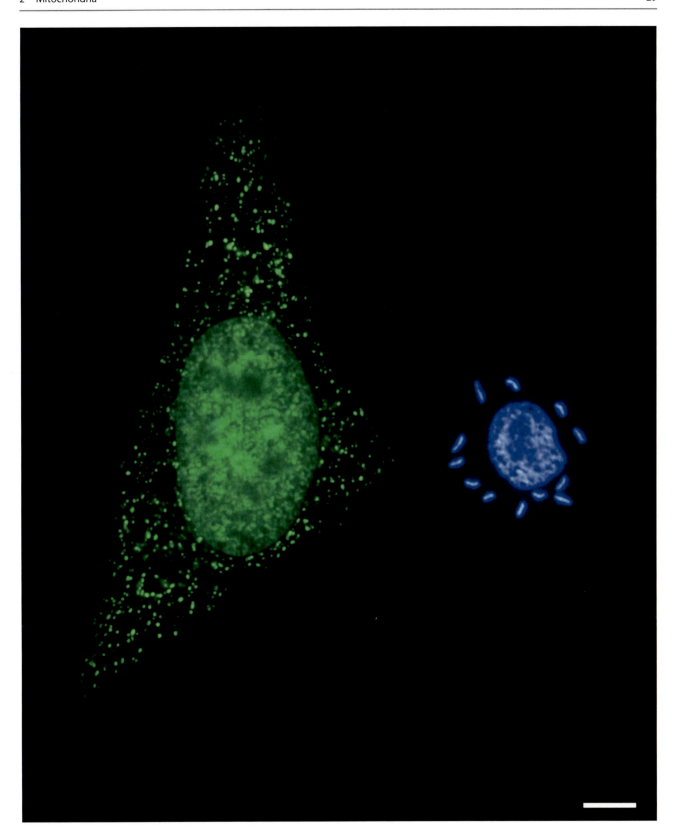

Plate 2.3

Uniparental inheritance of mitochondria during mating of *Didymium iridis*

The active, selective digestion of mitochondrial DNA (mtDNA) from one parent in the zygote is a possible molecular mechanism for uniparental inheritance of mitochondria, but direct evidence has been observed in very few species [10]. These photographs show the behavior of mitochondria and mtDNA during mating of the myxamoeba of the true slime mold *Didymium iridis* (**A–D**) [11]. To show selective digestion of mtDNA in the zygote, two myxamoebal strains of *D. iridis* were crossed and mitochondria and mtDNA changes over time were observed by phase-contrast microscopy using alkaline fixation and DAPI staining. Each myxamoeba of *D. iridis* contains about 30 mitochondria, and the zygote had about 60. Each mitochondrion contains rod-shaped mtDNA which is visualized as white fluorescence following DAPI staining (**B**). 4.5 h after mating, the fluorescence of mtDNA in about 30 mitochondria decreased to small fluorescent spots (**C**) which disappeared completely by 5 h after mating (arrow in white). By contrast, the mtDNA in the other 30 mitochondria as well as all of their mitochondrial sheaths remained unchanged (arrow in black). The rapid, selective disappearance of mtDNA observed in *D. iridis* is likely the result of selective digestion of mtDNA from one parent, as is observed in other known cases of mtDNA disappearance.

These photographs are merged phase-contrast and fluorescence microscopy images showing disappearance of mtDNA during zygote formation of *D. iridis* [11]. Two myxamoebal strains with different mating types were cultured on PGY plates at 23 °C with live bacteria (*Klebsiella aerogenes*) for food. Zygote formation was induced on mating plates at 23 °C. The two strains were mixed equally, and the suspensions were plated at 2,000 cells/mm^2. Myxamoebae and zygotes were fixed with small drops of formaldehyde solution adjusted to the alkalinity on the 1.5 % agar plate. DNA was stained with 4′,6-diamidino-2-phenylindole (DAPI) and a coverslip was placed over the stained sample. Photographs were taken with an epifluorescence microscope equipped with a CCD camera. Scale bars: 10 μm (**A**), 2 μm (**B–D**). This figure is adapted from [11].

Contributors

Yohsuke Moriyama[1], Shigeyuki Kawano[2]*, [1]Department of Anatomy II and Cell Biology, School of Medicine, Fujita Health University, 1-98 Dengakugakubo, Kutsukake-cho, Toyoake, Aichi 470-1192, Japan, [2]Department of Integrated Biosciences, Graduate School of Frontier Sciences, The University of Tokyo, Bldg. FSB-601, 5-1-5 Kashiwanoha, Kashiwa, Chiba 277-8562, Japan
*E-mail: kawano@k.u-tokyo.ac.jp

References

10. Moriyama Y, Kawano S (2003) Rapid, selective digestion of mitochondrial DNA in accordance with the *matA* hierarchy of multiallelic mating-types in the mitochondrial inheritance of *Physarum polycephalum*. Genetics 164:963–975
11. Moriyama Y, Itoh K, Nomura H, Kawano S (2009) Disappearance of mtDNA during mating of the true slime mold *Didymium iridis*. Cytologia 74:159–164

2 Mitochondria

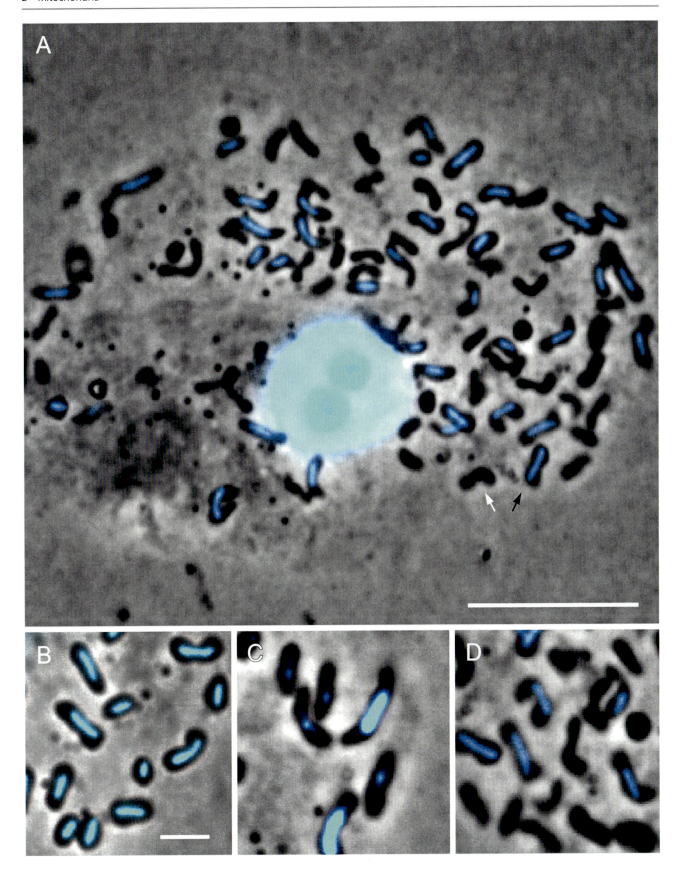

Plate 2.4

Giant mitochondrion in synchronized *Chlamydomonas* cells

Giant mitochondria of various shapes were formed by fusion of smaller and reticular mitochondria at early and intermediate stages of the growth phase of the cell cycle. Marked changes in mitochondrial morphology observed in synchronized cells mean dynamic changes of their membranes. Figures show 3D images of each cell containing reticular mitochondria (**A**), and a giant global mitochondrion (**B**). We found that the total surface areas of mitochondria in a cell dramatically decreases, while the total volume remaining constant, concurrent with the formation of a global giant mitochondria (**b**) from smaller reticular mitochondria (**a**), which returns to the initial level concomitant with the reversion of the giant to the reticular forms during the cell cycle. Giant mitochondrion formation causes temporary reduction of respiratory function, suggesting that these mitochondria play roles in the cell cycle other in addition to respiration. Giant mitochondrion formation and reduction of respiratory function may occur for the purpose of recombination, regulation, and information exchange between mitochondrial DNA by fusion of reticular and small independent mitochondria.

Behavior of mitochondria in *Chlamydomonas reinhardtii* cells was studied with a mitochondrial membrane binding fluorescent dye, 2-(p-Dimethylamino-styryl)-1-methyl-pyridine iodine (DASMPI), and 3D electron microscopy images [12–14]. *Chlamydomonas* cells were synchronized under a 12 h light : 12 h dark regimen at 25 °C under photoautotrophic conditions. Figure (**b**) shows the fluorescence microscopy profile of a DASMPI-stained giant global mitochondrion in a cell captured 1.5 h after the onset of the light period. A portion of culture was taken out and DASMPI was added to give a final concentration of 100 µg/mL. After 5 m, cells were washed with culture medium, and DASMPI-stained mitochondria were observed by fluorescence microscopy (yellow: mitochondria, red: chloroplast). For electron microscopy, cells were treated with glutaraldehyde at a final concentration of about 1 % (v/v) and placed at 4 °C for 2 h. Samples were then washed with 0.1 M phosphate buffer (pH 7.2). Cells were resuspended in 0.5–1 % (w/v) osmium tetroxide in water, then incubated at 4 °C for 1 h. The suspension was centrifuged, embedded in Spurr resin, and sectioned into 96 ultrathin slices on a Porter-Blum MT-1 microtome using a diamond knife. Sections were stained with 3 % uranyl acetate solution for washed with water, and stained with lead salt solution, after which they were examined with an electron microscope. These cell sections were photographed at an appropriate magnification, and organelles in each section were drawn on tracing paper. Data were entered into software using a digitizing table. Three-dimensional computer data were viewed on a color display and the image was rotated in space until a proper angle of view was obtained (yellow: mitochondria, red: nucleus, green: chloroplast, blue: cell wall). Scale bars: 2 µm. This figure is adapted from [13].

Contributors

Tomoko Ehara[1]*, Tetsuaki Osafune[2], [1]Department of Microbiology, Tokyo Medical University, Shinjuku, Tokyo 160-8402, Japan, [2]Beppubay Research Institute for Applied Microbiology, 5-5 Kajigahama, Kitsuki 873-0008, Japan
*E-mail: e-tomo@tokyo-med.ac.jp

References

12. Ehara T (1998) Studies of mitochondria during the cell cycle of *Chlamydomonas*. Plant Morphol 10:50–59 (Japanese)
13. Ehara T, Osafune T, Hase E (1995) Behavior of mitochondria in synchronized cells of *Chlamydomonas reinhardtii* (Chlorophyta). J Cell Sci 108:499–507
14. Osafune T (2006) Behavior of mitochondria, chloroplast and pyrenoid in synchronized cells of *Chlamydomonas* and *Euglena*. Plant Morphol 18:35–45 (Japanese)

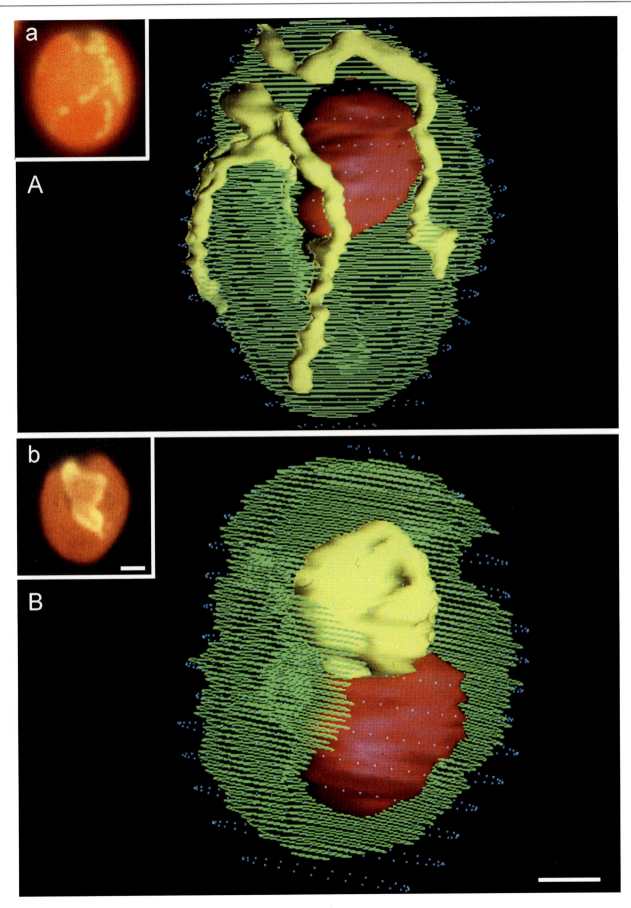

Plate 2.5

Dynamic transition of mitochondrial morphologies during germination in living zygospore

The unicellular green algae *Chlamydomonas reinhardtii* has a haploid life cycle, forms diploid zygotes, and has two mating types, mt+ and mt−. Vegetative cells have a network of string mitochondria, while zygote mitochondria change dynamically between periods of dormancy and meiotic division. These images show mitochondria (**A**) and its nuleoid (**B**) transition over the course of germination following light exposure (upper left; 0 h, upper middle; 6 h, upper right; 9 h, lower left; 12 h, lower middle; 13 h, lower right; 18 h) [15].

Mature zygotes (zygospores) responde to environmental signals such light and nutrients, and germinate to generate haploid progeny. Mitochondria are initially particle-like and scattered at the surface of cells, while not localizing to the cellular interior (upper left in **A**). Gradually, mitochondria transform in shape from particle-like to short and tubular, and became branch-like structures scattered throughout the cell (upper middle and right in **A**). Mitochondria begin to assemble in the area surrounding the nucleus (lower left **A**). After assembly, tubular mitochondria begin to migrate radically to the other hemisphere via the zygote surface layer (lower middle in **A**). As a result of meiotic division, four daughter cells, with tubular mitochondria, are formed (right lower in **A**). During this transition, microtubules are involved in mitochondrial accumulation around the cell nucleus, while microfilaments might maintain the tubular network of mitochondria [16]. Mitochondrial nucleoids were localized to mitochondria, and their form is particle-like. Nucleoids become long and stringy, divided, and form new particle-like nucleoids in zygospores 9 h after light exposure (upper right in **B**). Newly created nucleoids move in coordination with the mitochondrial transition. In the end, mitochondrial nucleoids are evenly distributed to daughter cells.

Cells of both mating types were cultured on agar plates with Snell's medium. For gamete induction, the two cell types were suspended separately in nitrogen-free solution and incubated for 3 h under light conditions. Equal volumes of the both gamete suspensions were mixed and spread on agar plates. Zygotes were incubated for 7 days in the dark and re-exposed to light to initiate germination. To stain cell nuclei and mitochondrial and chloroplast nucleoids in living zygotes, SYBR Green I was added to cell suspensions to a final dilution of 1:1,000. To observe the morphology of mitochondria in living zygotes, cells were stained with $DiOC_6$ at a final concentration of 0.1 µg/mL. The mixtures of cells with dyes were kept in the dark for 30 min and washed by gentle centrifugation. Cells were observed with a fluorescence microscope with excitation at 488 nm. Scale bar: 5 µm. This figure is adapted from [16].

Contributors

Hiroaki Aoyama[1,2], Soichi Nakamura[2]*, [1]Center of Molecular Biosciences, Tropical Biosphere Research Center, [2]Laboratory of Cell and Functional Biology, Faculty of Science, University of the Ryukyus, Nishihara, Okinawa 903–0213, Japan
*E-mail: nsoichi@sci.u-ryukyu.ac.jp

References

15. Aoyama H, Hagiwara Y, Misumi O, Kuroiwa T, Nakamura S (2006) Complete elimination of maternal mitochondrial DNA during meiosis resulting in the paternal inheritance of the mitochondrial genome in *Chlamydomonas* species. Protoplasma 228:231–242
16. Aoyama H, Kuroiwa T, Nakamura S (2009) The dynamic behavior of mitochondria in living zygotes during maturation and meiosis in *Chlamydomonas reinhardtii*. Eur J Phycol 44:497–507

2 Mitochondria

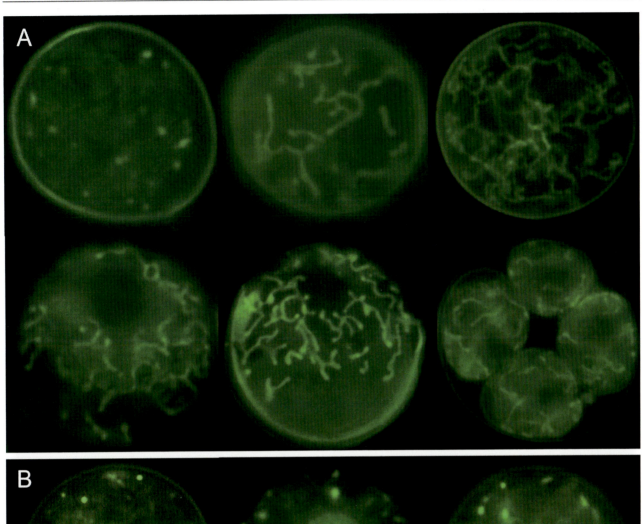

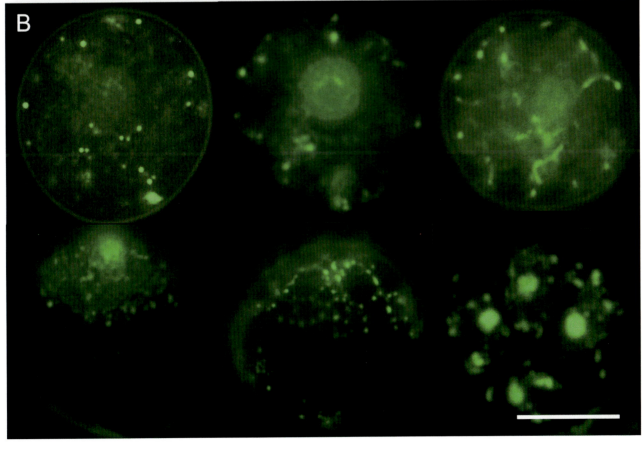

Plate 2.6

Mitochondrial nucleoids in the *Euglena gracilis* mitochondrial network

The mitochondrial network is situated close to the cell surface, surrounding other organelles in the cell. The volume of the cell gradually increases during the growth phase and reaches a maximum at the initial stage of mitosis. During interphase, about 75 % of the mitochondrial nucleoids were uniform, and the remaining ones were larger. Elongated nucleoids and pairs of nucleoids situated close to one another were always observed in the non-dividing growing cells. The mitochondrial network is maintained during cellular cleavage. In the cells undergoing division about 300 mitochondrial nucleoids could be counted before cytokinesis. During cytokinesis, about half of the mitochondrial nucleoids are distributed to each of the two daughter cells. Shortly after cell division, daughter cells contain about 150 mitochondrial nucleoids. Nucleoids in the daughter cells increase in number as cell increase in size, with number of nucleoids reaching double the initial value just prior to mitosis, while the total length of the mitochondrial network is always proportional to the cell volume. DNA in the nucleoids of mitochondria was detected with anti-DNA antibody and immunoelectron microscopy. Most nucleoids in mitochondria were 70–130 nm in diameter. These contained a core in which DNA threads were in tight contact [17]. Nucleiod structure was very different from those previously observed by conventional electron microscopy.

The figure shows a fluorescence micrograph of *Euglena* cell double-stained with DiOC6(3) (3,3′-dehexyloxacarbocyanine iodide) and ethidium bromide [18]. Mitochondria are indicated by green fluorescence and the nucleus is indicated by red fluorescence. There were many small granules in the network of mitochondria. These granules appeared to be yellow, as a result of combination of red DNA stain and mitochondrial green. This *Euglena* cells do not have chloroplasts, as this was the UV-induced leuco mutant strain. The behavior and appearance of mitochondria in green *Euglena* were the same as those in the leuco *Euglena*.

DiOC6(3) was dissolved in ethanol at concentration of 1 % and diluted in culture medium to a concentration of 5 µg/mL. Cells were treated with DiOC6(3) for 5 min at room temperature. After a brief rinse with culture medium, cells were further treated for 5 min with 0.2 % ethidium bromide, washed with culture medium, and examined under a fluorescence microscope with a blue excitation filter. Scale bar: 10 µm. This figure is adapted from [18].

Contributors

Yasuko Hayashi[1]*, Katsumi Ueda[2], [1]Department of Environmental Science, Graduate School of Science and Technology, Niigata University, Ikarashi, Niigata 950-2181, Japan, [2]Deceased
*E-mail: yhayashi@env.sc.niigata-u.ac.jp

References

17. Hayashi-Ishimaru Y, Ueda K, Nonaka M (1993) Detection of DNA in the nucleoids of chloroplasts and mitochondria in Euglena gracilis by immunoelectron microscopy. J Cell Sci 106:1159–1164

18. Hayashi Y, Ueda K (1989) The shape of mitochondria and the number of mitochondrial nucleoides during the cell cycle of *Euglena gracilis*. J Cell Sci 93:565–570

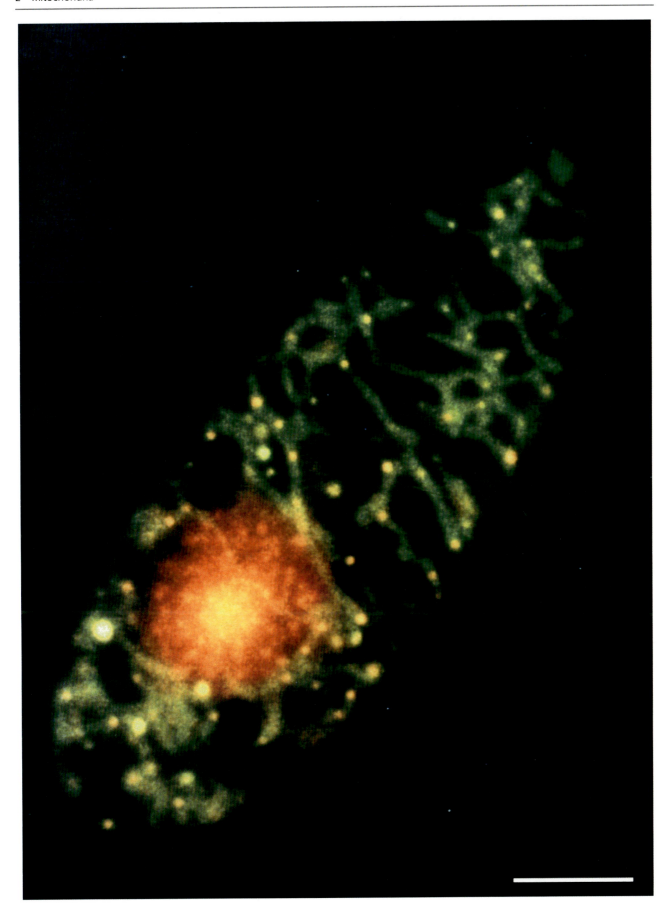

Plate 2.7

Mitochondrial fission and fusion in an onion epidermal cell

Mitochondria in an onion epidermal cell are highlighted by the fluorescent protein, Kaede, localized to mitochondrial matrix (**top panel**). The cell has more than 10,000 small mitochondria particles (green dots). The right portion of the cell was exposed to 400 nm light to irreversibly change the color of Kaede from green to red. The second panel shows the cell 1 min after photo-conversion, resulting in co-existence of green and red mitochondria. Some red mitochondria in the left portion and some green mitochondria in the right are the results of movement from the opposite sides of the cell 1 and 5 min after photo-conversion (**second** and **third panels**, respectively). Fusion of green and red mitochondria results in yellow mitochondria (the bottom panels show the time-course images of mitochondrial fusion). Two hours after photo-conversion mitochondria are almost uniformly yellow (**fourth panel**). Balanced fission and fusion makes exchange of protein among all mitochondria in the cell possible without changing the particulate shape of mitochondria.

Onions were purchased from a grocery store. The bulb epidermis was peeled and used for transformation and observation. *Kaede* ORF with mitochondrial the Arabidopsis ATPase delta prime subunit presequence was cloned into an expression plasmid with CaMV35S promoter and NOS terminator. The plasmid was introduced into onion epidermal cells by particle bombardment according to the manufacturer's instructions. Images were taken with a fluorescent microscope with confocal laser scanning unit. Excitation lasers were 488 nm and 561 nm for green and red fluorescence, respectively. Additional details of the experiment to observe mitochondrial fission and fusion in plants and their movies are available at the website for the paper [19]. Scale bars: 10 μm (**main**), 1 μm (**inset**). This figure is adapted from [19].

Contributors

Shin-ichi Arimura*, Graduate School of Agricultural and Life Sciences, The University of Tokyo, 1-1-1 Yayoi, Bunkyo-ku, Tokyo 113-8657, Japan
*E-mail: arimura@mail.ecc.u-tokyo.ac.jp

References

19. Arimura S, Yamamoto J, Aida GP, Nakazono M, Tsutsumi N (2004) Frequent fusion and fission of plant mitochondria with unequal nucleoid distribution. Proc Natl Acad Sci U S A 101:7805–7808

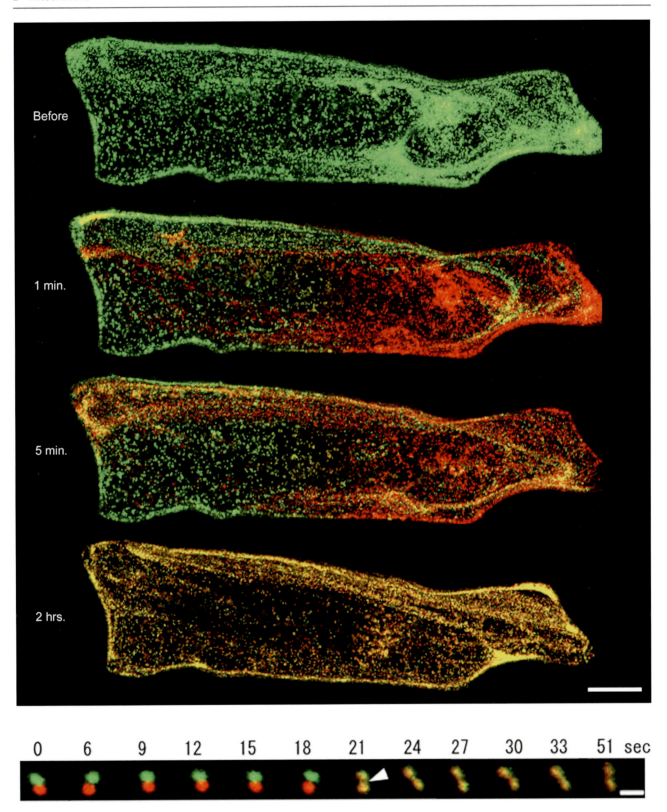

Plate 2.8

Mitochondria in *Arabidopsis* guard cells

Micrographs of mitochondria are presented side-by-side with live cell imaging of cells expressing a mitochondrial GFP marker (**B**) and transmission electron microscopy using metal-contact type rapid freezing method with liquid helium (**A**). Both pictures are of Arabidopsis cotyledon guard cells. Most mitochondria are spherical in shape with diameter ranging from 0.5 to 1.0 μm.

Seeds of *Arabidopsis thaliana* (Col) were sterilized and sown on MS medium, vernalized in the dark at 4 °C for 3–4 days, then grown at 23 °C under continuous light. For transmission electron microscopy, cotyledon of TG2-1 (γ-TIP-GFP expressing line) were rapidly frozen by slamming onto the copper block cooled by liquid helium, using a HIF4K automated device [20]. Frozen samples were then transferred to 4 % OsO_4 in anhydrous acetone and kept at −80 °C for 4 days. Samples were held at −20 °C for 2 h, 4 °C for 2 h, then at room temperature for 10 m, then washed with anhydrous acetone. Samples were embedded in Spurr resin. Ultrathin sections were cut, stained and observed under a transmission electron microscope. For live cell imaging of mitochondria with GFP-marker, cotyledon of MT42 (coxIV-GFP expressing line, generous gift of seeds from Prof. Maureen Hanson, Cornell University) plants were mounted on glass slides with MS medium and covered with cover slips, then observed by confocal laser scanning microscope. Scale bars: 10 μm.

Contributors

Chieko Saito*, Department of Biological Sciences, Graduate School of Science, The University of Tokyo, 7-3-1 Hongo, Bunkyo-ku, Tokyo 113-0033, Japan
*E-mail: chiezo@bs.s.u-tokyo.ac.jp

References

20. Saito C (2013) Technical note: an electron micrograph of guard cells cryofixed by metal contact type rapid freezing method. Cytologia 78:1–2

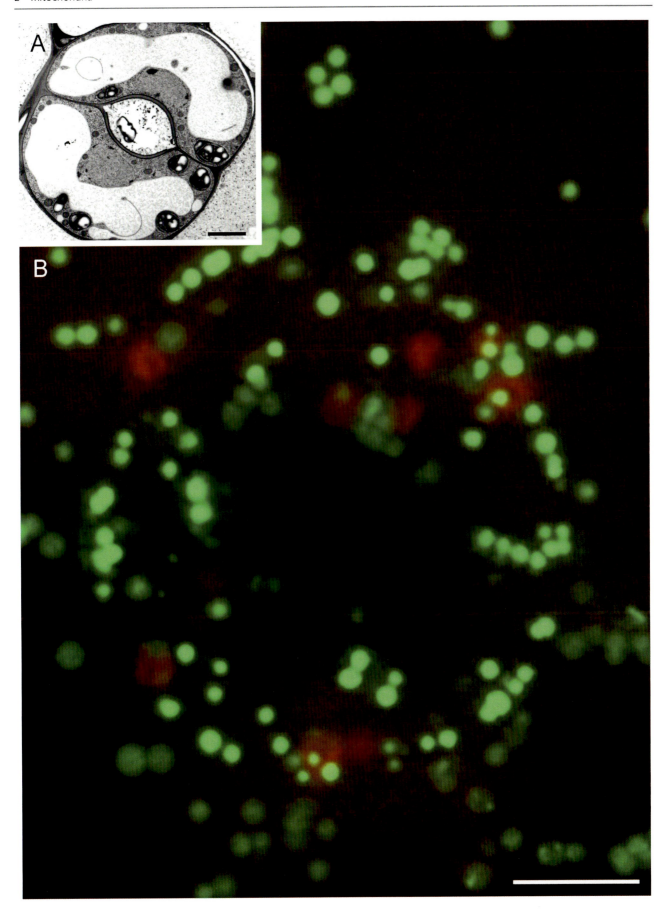

Plate 2.9

Mitochondria of thermogenic skunk cabbage

Sex-dependent thermogenesis during reproductive organ development in the inflorescence is a characteristic feature of some protogynous arum species [21]. One of such plants, skunk cabbage (*Symplocarpus renifolius*, **B**), produces massive heat (around 20 °C) during the female stage but does not produce it during the subsequent male stage in which the stamen completes development, the anthers split open, and pollen is released. Recent studies have identified the spadix as the thermogenic organ [21, 22].

Bisexual flowers of skunk cabbage are comprised a large number of florets arranged on the surface of the spadix [21]. Each flower is composed of one pistil surrounded by four stamens and four petals. During the female stage, the stamens display extensive structural rearrangements including changes in organelle structure and density. Stamens accumulate high levels of mitochondrial proteins, including possible thermogenic factors, alternative oxidase, and uncoupling protein. The petals and pistils do not exhibit extensive changes during the female stage. However, they contain a larger number of mitochondria during the male stage in which they develop large cytoplasmic vacuoles. Comparison between female and male spadices shows that mitochondrial number rather than their level of activity is relevant to thermogenesis. Spadices, even in the male, contain a larger numbers of mitochondria and consume more oxygen when compared to non-thermogenic plants.

Female stage petals were observed by transmission electron microscopy (**A, C**) [21]. Samples were dissected using clamping forceps. Tissues were fixed with 4 % paraformaldehyde and 2 % glutaraldehyde in 100 mM sodium cacodylate buffer (pH 7.4) for 2 h at room temperature and postfixed with 1 % osmium tetroxide in 100 mM cacodylate buffer (pH 7.4) for 4 h at room temperature. After dehydration in a graded methanol series, tissue pieces were further dehydrated in methanol/propylene oxide (1:1 v/v), 100 % propylene oxide, and propylene oxide/Epon812 resin (1:1 v/v), and finally embedded in 100 % Epon812 resin. Ultrathin sectioning (70 nm) were carried out with a diamond knife on an ultramicrotome, and transferred to Formvar-coated grids. Sections were stained with 4 % uranyl acetate for 12 min then with lead citrate solution for 3 min and examined with a transmission electron microscope at 80 kV. Images were acquired using a CCD camera. Mitochondria (M), amyloplasts (Am), nucleus (N), cell wall (CW). Scale bars: 1 μm. This figure is adapted from [21, 22].

Contributors

Mayuko Sato[1]*, Yasuko Ito-Inaba[2], [1]RIKEN Center for Sustainable Resource Science, 1-7-22 Suehiro-cho, Tsurumi-ku, Yokohama, Kanagawa 230-0045, Japan, [2]Organization for Promotion of Tenure Track, University of Miyazaki, 1-1 Gakuenkibanadai-nishi, Miyazaki 889-2192, Japan
*E-mail: mayuko.sato@riken.jp

References

21. Ito-Inaba Y, Sato M, Masuko H, Hida Y, Toyooka K, Watanabe M, Inaba T (2009) Developmental changes and organelle biogenesis in the reproductive organs of thermogenic skunk cabbage (*Symplocarpus renifolius*). J Exp Bot 60:3909–3922
22. Ito-Inaba Y, Hida Y, Mori H, Inaba T (2008) Molecular identity of uncoupling proteins in thermogenic skunk cabbage. Plant Cell Physiol 49:1911–1916

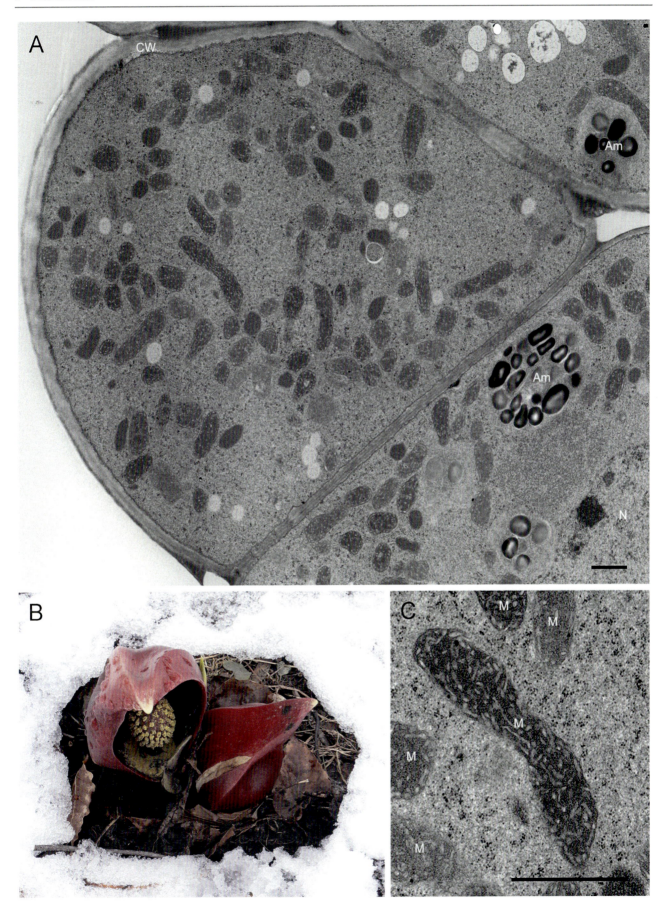

Chapter References

1. Kuroiwa T (1982) Mitochondrial nuclei. Int Rev Cytol 75:1–59
2. Miyakawa N, Aoi H, Sando N, Kuroiwa T (1991) Mitochondrial import proteins. Nature 349(6306)
3. Kuroiwa T, Miyakawa I (1991) In: Broach JR et al (eds) The molecular and cellular biology of the yeast *Saccharomyces*. Cold Spring Harbor Laboratory Press, New York
4. Kuroiwa T, Kojima H, Miyakawa I, Sando N (1984) Meiotic karyotype of the yeast *Saccharomyces cerevisiae*. Exp Cell Res 153:259–265
5. Kuroiwa T, Miyamura S, Kawano S, Hizume M, Toh-e A, Miyakawa I, Sando N (1986) Cytological characterization of a synaptonemal complex-less nucleolar-organizing region on a bivalent in the bivalent of *Saccharomyces cerevisiae* using a video-intensified microscope system. Exp Cell Res 165:199–206
6. Watson JD, Hopkins NH, Roberts JW, Steitz JA, Weiner AM (1987) Molecular biology of the gene, 4th edn. Benjamin Cummings, San Francisco, pp 552–553
7. Kuroiwa T, Kawano S, Nishibayashi S, Sato C (1982) Maternal inheritance of chloroplast DNA. Nature 298(5873):481–483
8. Kuroiwa T (2010) Review of cytological studies on cellular and molecular mechanisms of uniparental (maternal or paternal) inheritance of plastid and mitochondrial genomes induced by active digestion of organelle nuclei (nucleoids). J Plant Res 123:207–230
9. Ozawa S, Sasaki N (2009) Visualization of nucleoids in living human cells using SYBR Green I. Cytologia 74:365–366
10. Moriyama Y, Kawano S (2003) Rapid, selective digestion of mitochondrial DNA in accordance with the matA hierarchy of multiallelic mating-types in the mitochondrial inheritance of *Physarum polycephalum*. Genetics 164:963–975
11. Moriyama Y, Itoh K, Nomura H, Kawano S (2009) Disappearance of mtDNA during mating of the true slime mold *Didymium iridis*. Cytologia 74:159–164
12. Ehara T (1998) Studies of mitochondria during the cell cycle of *Chlamydomonas*. Plant Morphol 10:50–59 (Japanese)
13. Ehara T, Osafune T, Hase E (1995) Behavior of mitochondria in synchronized cells of *Chlamydomonas reinhardtii* (Chlorophyta). J Cell Sci 108:499–507
14. Osafune T (2006) Behavior of mitochondria, chloroplast and pyrenoid in synchronized cells of *Chlamydomonas* and *Euglena*. Plant Morphol 18:35–45 (Japanese)
15. Aoyama H, Hagiwara Y, Misumi O, Kuroiwa T, Nakamura S (2006) Complete elimination of maternal mitochondrial DNA during meiosis resulting in the paternal inheritance of the mitochondrial genome in *Chlamydomonas* species. Protoplasma 228:231–242
16. Aoyama H, Kuroiwa T, Nakamura S (2009) The dynamic behavior of mitochondria in living zygotes during maturation and meiosis in *Chlamydomonas reinhardtii*. Eur J Phycol 44:497–507
17. Hayashi-Ishimaru Y, Ueda K, Nonaka M (1993) Detection of DNA in the nucleoids of chloroplasts and mitochondria in *Euglena gracilis* by immunoelectron microscopy. J Cell Sci 106:1159–1164
18. Hayashi Y, Ueda K (1989) The shape of mitochondria and the number of mitochondrial nucleoides during the cell cycle of *Euglena gracilis*. J Cell Sci 93:565–570
19. Arimura S, Yamamoto J, Aida GP, Nakazono M, Tsutsumi N (2004) Frequent fusion and fission of plant mitochondria with unequal nucleoid distribution. Proc Natl Acad Sci USA 101:7805–7808
20. Saito C (2013) Technical note: an electron micrograph of guard cells cryofixed by metal contact type rapid freezing method. Cytologia 78:1–2
21. Ito-Inaba Y, Sato M, Masuko H, Hida Y, Toyooka K, Watanabe M, Inaba T (2009) Developmental changes and organelle biogenesis in the reproductive organs of thermogenic skunk cabbage (*Symplocarpus renifolius*). J Exp Bot 60:3909–3922
22. Ito-Inaba Y, Hida Y, Mori H, Inaba T (2008) Molecular identity of uncoupling proteins in thermogenic skunk cabbage. Plant Cell Physiol 49:1911–1916

Chloroplasts

3

Shigeyuki Kawano

Chloroplasts are photosynthetic organelles found in plant cells and eukaryotic algae. They absorb sunlight and use it in conjunction with water and carbon dioxide gas to produce starch. Photosynthetic pigments such as chlorophylls capture the energy from sunlight, and RuBisCO (ribulose-1,5-bisphosphate carboxylase/oxygenase) fixes carbon dioxide as a carbon source. Chloroplasts are not only the photosynthetic organelle but also the site of synthesis of many other important compounds such as pigments, fatty acids and amino acids. Chloroplasts, which are by definition plastids containing chlorophyll, and other plastids, such as etioplasts, leucoplasts, amyloplasts, and chromoplasts, develop either by division of an existing plastid or from proplastids. Proplastids arise during germ cell formation.

Chloroplasts are generally believed to have originated as endosymbiotic cyanobacteria. In this respect they are similar to mitochondria, but are found only in plants and photosynthetic protists. Both organelles are surrounded by a double membrane; both have their own DNA and are involved in energy metabolism. There are two main types of plastids depending on their membrane structure: primary plastids and secondary plastids. Primary plastids are found in glaucophytes, red, and green algae, including land plants, and secondary plastids are found in chlorophyll *c*-possessing algae, euglenophytes and chlorarachniophytes. Exploring the origin of plastids provide an insight into our understanding of the basis of photosynthesis in green plants, our primary food source.

In this chapter, various kinds of chloroplasts in land plants and algal cells will be highlighted, and related apparatuses will be illustrated. In the first three articles, S. Miyagishima et al., Y. Yoshida et al., and H. Hashimoto et al. review chloroplast division machinery in the unicellular red alga, *Cyanidioschyzon merolae* and the glaucocystophyte alga, *Cyanophora paradoxa*. Pyrenoids are sub-cellular compartments found in many algal chloroplasts, and their main function is to act as centers of carbon dioxide fixation in which ribulose-1,5-bisphosphate carboxylase/oxygenase (RuBisCO) is thought to be accumulated. T. Osafune et al. demonstrate the distribution of RuBisCO in synchronized *Euglena* cells by immunoelectron microscopy, showing RuBisCO accumulates in the pyrenoid. *Haematococcus pluvialis* is a freshwater species of green algae which is well known for its accumulation of carotenoids (e.g., astaxanthin) during encystment. S. Ota et al. present ultrastructural 3D reconstructions based on over 350 serial sections per cell to visualize the dynamics of astaxanthin accumulation and sub-cellular changes during encystment.

Many lower land plants (i.e., archegoniate plants) have cells containing only a single chloroplast. M. Shimamura et al. present monoplastidic cells of the hornwort, *Anthoceros punctatus*, and a liverwort, *Blasia pusilla*. Unique aquatic angiosperm *Podostemaceae* plants (riverweeds) have two differently sized chloroplasts in each epidermal cell. R. Fujinami presents dimorphic chloroplasts in the epidermis of *Podostemaceae* plants. A. Kondo et al. and C. Saito describe the distribution of chloroplasts and mitochondria in mesophyll cells of the flowering plant, *Kalanchoë blossfeldiana* and in *Sorghum* (a genus of numerous species of grasses). Y. Hayashi presents prolamellar bodies of the etioplast in etiolated cotyledon in the model land plant, *Arabidopsis thaliana*. Finally, H. Kuroiwa and T. Kuroiwa present the chloroplast division machinery of *Pelargonium zonale*, and Y. Nishimura presents active digestion of paternal chloroplast DNA in a young zygote of a model green alga, *Chlamydomonas reinhardtii*.

S. Kawano (✉)
Department of Integrated Biosciences, Graduate School of Frontier Sciences, The University of Tokyo, 5-1-5 Kashiwanoha, Kashiwa, Chiba 277-8562, Japan
e-mail: kawano@k.u-tokyo.ac.jp

T. Noguchi et al. (eds.), *Atlas of Plant Cell Structure*,
DOI 10.1007/978-4-431-54941-3_3, © Springer Japan 2014

Plate 3.1

Chloroplast division by the plastid-dividing ring

Plastids evolved from a cyanobacterial endosymbiont and their continuity is maintained by plastid division and segregation which is regulated by the eukaryotic host cell. Plastids divide by constriction of the inner and outer envelope membranes [1]. Ring-like structures called plastid-dividing rings have been identified in the red alga *Cyanidium caldarium* by transmission electron microscopy [2], and, since their discovery, plastid-dividing rings have been identified in several lineages of photosynthetic eukaryotes including land plants. Based on these observations, recent studies have identified several components of the plastid division machinery. The division complex has retained certain components of the cyanobacterial division complex along with components developed by the host cell [1]. Based on the molecular components of the division complex which have been identified, it is becoming increasingly clear how the division complex has evolved and how it is assembled, constricted, and regulated in the host cell.

These images show a dividing chloroplast of the unicellular red alga *Cyanidioschyzon merolae* as visualized by transmission electron microscope (**A**, TEM) and a field emission scanning electron microscope (**B**, FE-SEM). For TEM, cell and chloroplast division of *C. merolae* were synchronized by exposing cells to a light/dark cycle. Synchronized cells were rapidly frozen in liquid propane (-195 ℃) and fixed with 1 % OsO_4, then dissolved in acetone at -80 ℃. After samples were warmed gradually to room temperature, they were embedded in Spurr resin. Serial thin sections (each 70 nm thick) were stained with uranyl acetate and lead citrate, and examined with TEM [3]. For FE-SEM, dividing chloroplasts were isolated from synchronized *C. merolae* culture and were fixed with 1 % glutaraldehyde. After dehydration, chloroplasts were dried to the critical point, and then mounted and sputter-coated with platinum. The samples were examined with FE-SEM [4]. Scale bars: 0.5 μm. This figure is adapted from [4].

Contributors

Shin-ya Miyagishima[1]*, Tsuneyoshi Kuroiwa[2], [1]Center for Frontier Research, National Institute of Genetics, 1111 Yata, Mishima, Shizuoka 411-8540, Japan, [2]CREST, Initiative Research Unit, College of Science, Rikkyo University, Toshima, Tokyo 171-8501, Japan
*E-mail: smiyagis@nig.ac.jp

References

1. Miyagishima S, Kabeya Y (2010) Chloroplast division: squeezing the photosynthetic captive. Curr Opin Microbiol 13:738–746
2. Mita T, Kanbe T, Tanaka K, Kuroiwa T (1986) A ring structure around the dividing plane of the *Cyanidium caldarium* chloroplast. Protoplasma 130:211–213
3. Miyagishima S, Nishida K, Mori T, Matsuzaki M, Higashiyama T, Kuroiwa H, Kuroiwa T (2003) A plant-specific dynamin-related protein forms a ring at the chloroplast division site. Plant Cell 15:655–665
4. Miyagishima S, Itoh R, Aita S, Kuroiwa H, Kuroiwa T (1999) Isolation of dividing chloroplasts with intact plastid-dividing rings from a synchronous culture of the unicellular red alga *Cyanidioschyzon merolae*. Planta 209:371–375

3 Chloroplasts

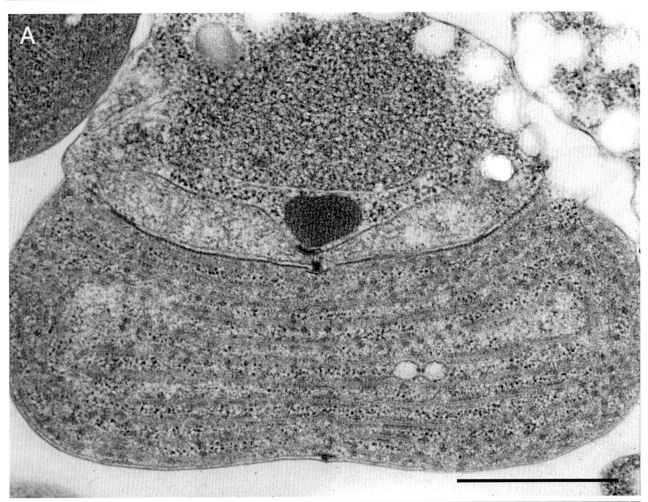

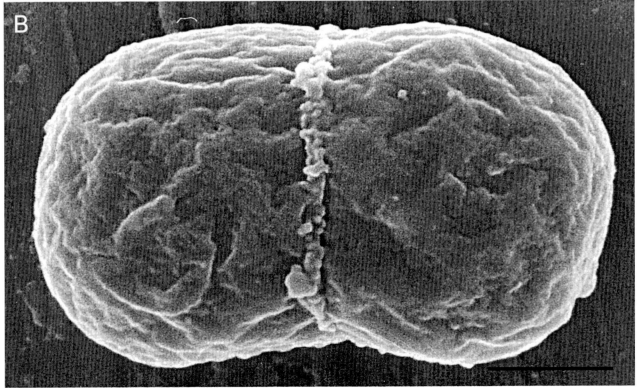

Plate 3.2

Chloroplasts divide by contraction of a bundle of polyglucan nanofilaments

Plastids such as chloroplasts arose from a cyanobacterial endosymiont and have retained their own genome. Consistent with their bacterial origin, plastids multiply by binary division of pre-existing organelles, which is executed by a complex called the plastid dividing (PD) machinery [5]. The glycosyltransferase protein plastid-dividing ring 1 (PDR1) was identified by proteomic analysis of PD machinery isolated from unicellular *Cyanidioschyzon merolae*, which contains a single chloroplast and a single mitochondrion [5]. Together with carbohydrates the PDR1 protein forms a ring that constricts to physically divide the plastid. Figures of A and B show an immuno-electron (EM) micrograph (**A**) and an immunofluorescence image (**B**) of isolated PD machinery. Immunogold particles indicating PDR1 (black dots) are localized throughout each of the PD machineries. Many more immunogold particles appear in the less condensed PD ring filament region than in the solid PD ring filament region, suggesting that PDR1 proteins are associated with the whole of the PD ring, from the inside to the outside. Fluorometric saccharide ingredient analysis of purified PD ring filaments showed that only glucose was included. Thus, the PD ring is made up of polysaccharide chains and proteins, which together generate a ring that constricts to divide the plastids. The EM image was chosen as the cover of *Science* (Vol. 329, no. 5994). Recently, it was revealed that mitochondrial and plastid divisions are regulated by a kinesin-like protein TOP (green) (**C**) [6]. In the early phase of division, TOP promotes Aurora kinase localization to activate division machineries by protein phosphorylation. A series of studies have uncovered important and unexpected cooperative behaviors of organelle division machineries and cell proliferation mechanisms.

In order to isolate intact PD machineries, dividing plastids were isolated from synchronized *C. merolae* cells at M phase. Isolated PD machineries form not only ring but also spiral and supertwist structures (**D**) [7]. The immunofluorescence image (**D**) was chosen as the highlight of *Science* (Vol. 313, no. 5792). In figure B, an isolated PD machinery was immunostained with PDR1 (green), Dnm2 (red) and FtsZ2 (blue). For immuno-EM, samples were negatively stained with 0.5 % phosphotungstic acid (pH 7.0) and examined with an electron microscope. Scale bars: 0.1 μm (**A**), 1 μm (**B**, **C**). This figure is adapted from [5, 6, 7].

Contributors

Yamato Yoshida[1,2], Haruko Kuroiwa[1], Tsuneyoshi Kuroiwa[1]*, [1]CREST, Initiative Research Unit, College of Science, Rikkyo University, Toshima, Tokyo 171-8501, Japan, [2]Present address: Department of Plant Biology, Michigan State University, East Lansing, MI 48824-1312, USA
*E-mail: tsune@rikkyo.ne.jp

References

5. Yoshida Y, Kuroiwa H, Misumi O, Yoshida M, Ohnuma M, Fujiwara T, Yagisawa F, Hirooka S, Imoto Y, Matsushita K, Kawano S, Kuroiwa T (2010) Chloroplasts divide by contraction of a bundle of nanofilaments consisting of polyglucan. Science 329:949–953
6. Yoshida Y, Fujiwara T, Imoto Y, Yoshida M, Ohnuma M, Hirooka S, Misumi O, Kuroiwa H, Kato S, Matsunaga S, Kuroiwa T (2013) The kinesin-like protein TOP promotes Aurora localization and induces mitochondrial, chloroplast and nuclear division. J Cell Sci 126:2392–2400
7. Yoshida Y, Kuroiwa H, Misumi O, Nishida K, Yagisawa F, Fujiwara T, Nanamiya H, Kawamura F, Kuroiwa T (2006) Isolated chloroplast division machinery can actively constrict after stretching. Science 313:1435–1438

3 Chloroplasts

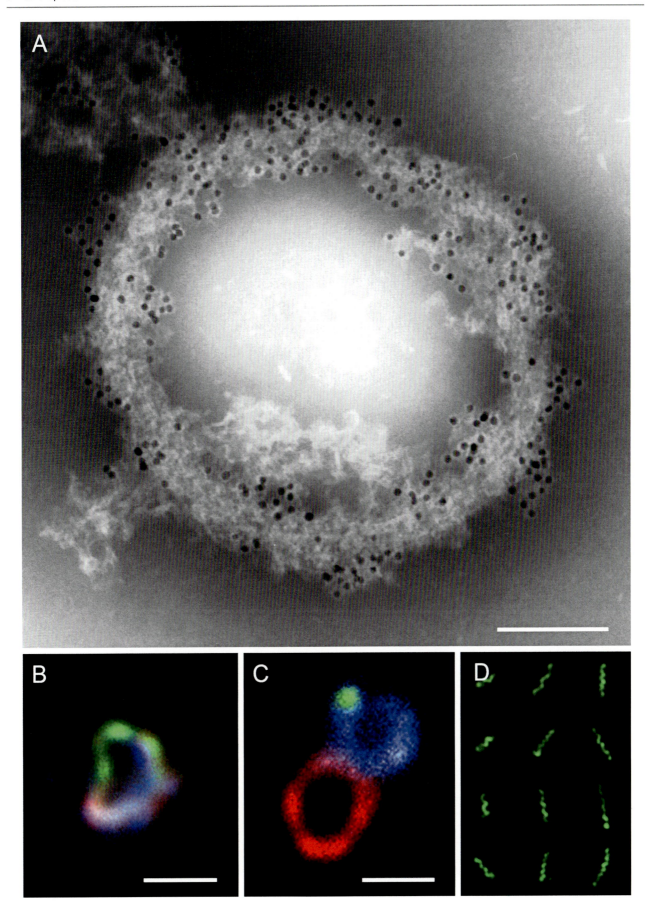

Plate 3.3

Cyanelle division of the glaucocystophyte alga *Cyanophora paradoxa*

Glaucocystophyte plastids are called cyanelles because they are surrounded by a peptidoglycan layer, resembling cyanobacterial cells, which are the putative ancestor of plastids. This suggests that cyanelles may be the most primitive among the known plastids, and this idea is supported by the sequence phylogeny of cyanelle DNA.

Cyanelles divide by binary fission as do other plastids. However, electron microscopic observations reveal that cyanelle division differs from typical plastid fission. Cyanelle division involves septum formation at the cleavage site by ingrowth of the peptidoglycan layer between the outer and inner envelope membranes. Electron-opaque rings are detected only on the stromal face of the inner envelope membrane (**C**, white arrow) but not on the cytoplasmic face of the outer envelope membrane (**C**) [8]. SEM images also confirm the absence of a cytoplasmic ring (**A**) [9]. As in cyanobacterial cell division, septum ingrowth may in part provide the external fission force, whereas in plastids without peptidoglycan the fission force may be provided by the cytoplasmic plastid-division ring. On the other hand, fluorescence immunostaining shows evidence for the presence of a FtsZ ring at the cleavage site (**B**, green fluorescence; white and red represent fluorescence of DAPI-stained nucleoids and chlorophyll autofluorescence, respectively) [10], a homologue of the bacterial cell division ring which is commonly present in plastids of other plants. These observations suggest that cyanelle division represents an intermediate stage between cyanobacterial and plastid division, a potential "missing link." If plastids have a monophyletic origin, the stromal ring, i.e. cyanelle ring, and homologous inner plastid-dividing ring in other plastids might have evolved before than the outer cytoplasmic plastid-dividing ring.

Cyanophora paradoxa Korshikov was obtained from strain NIES-547 of the algal collection of the National Institute of Environmental Studies (Tsukuba, Japan). Cells were axenically grown in C-medium at 25 °C under a diurnal 13:11 h light:dark regime. Cultures were illuminated with white fluorescent lamps at an intensity of approximately 50 μmol photons m-2 s-1. *C. paradoxa* cells were harvested by centrifugation at 1,000 g at room temperature and then fixed with 2.5 % (w/v) glutaraldehyde in 0.1 M sodium cacodylate buffer (pH 7.2) on ice for 2 h. They were washed with the same buffer several times and post-fixed with 1 % (w/v) osmium tetroxide in the same buffer on ice for 2 h. Thereafter, cells were rinsed with the same buffer several times and dehydrated through a graded acetone series on ice.

Samples were then infiltrated and embedded in Spurr resin. Ultrathin sections with grey–silver interference color were stained with uranyl acetate and lead citrate, then observed under a transmission electron microscope. Scale bars: 0.2 μm. This figure is adapted from [9, 10].

Contributors

Haruki Hashimoto[1]*, Mayuko Sato[2], Shigeyuki Kawano[3], [1]Department of Biological Sciences, Faculty of Science, Kanagawa University, 2946 Tsuchiya, Hiratsuka 259-1293, Japan, [2]RIKEN Center for Sustainable Resource Science, Tsurumi-ku, Yokohama 230-0045, Japan, [3]Department of Integrated Biosciences, Graduate School of Frontier Sciences, The University of Tokyo, Bldg. FSB-601, 5-1-5 Kashiwanoha, Kashiwa, Chiba 277-8562, Japan
*E-mail: hashimoto-h-bio@kanagawa-u.ac.jp

References

8. Iino M, Hashimoto H (2003) Intermediate features of cyanelle division of *Cyanophora paradoxa* (Glaucocystophyta) between cyanobacterial and plastid division. J Phycol 39:561–569
9. Sato M, Mogi Y, Nishikawa T, Miyamura S, Nagumo T, Kawano S (2009) The dynamic surface of dividing cyanelles and ultrastructure of the region directly below the surface in *Cyanophora paradoxa*. Planta 229:781–791
10. Sato M, Nishikawa T, Kajitani H, Kawano S (2007) Conserved relationship between FtsZ and peptidoglycan in the cyanelles of *Cyanophora paradoxa* similar to that in bacterial cell division. Planta 227:177–187

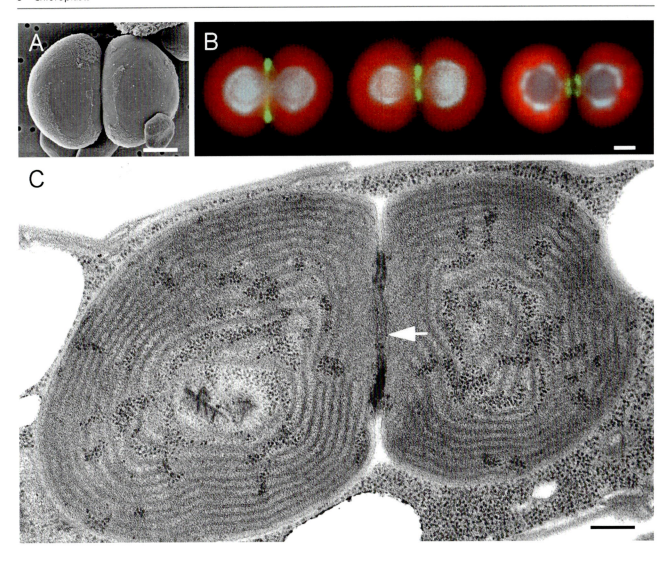

Plate 3.4

3D distribution of RuBisCO in synchronized *Euglena* cells

The pyrenoid exists in the chloroplasts of most eukaryotic algae and moss plants, and has been thought a mere storage place of RuBisCO (ribulose-1,5-bisphosphate carboxylase/oxygenase), though its actual functions are poorly understood.

Euglena gracilis cells were synchronized under a 14 h light and 10 h dark regimen at 25 °C under photoautotrophic conditions. Cell grew during the light period and divided during the following 10 h period, whether cells were in the dark or in the light [11]. Changes in pyrenoid morphology and the distribution of RuBisCO in the *Euglena* chloroplast were observed by immunoelectron microscopy [12] at different phases of the cell cycle in synchronized culture under photoautotrophic conditions. Immunoreactive proteins were concentrated in the pyrenoid (**A**) [13], and less densely distributed in the stroma during the light period, and gold particles are localized in the pyrenoid during the light period, which disappeared during the dark period with RuBisCO dispersing throughout the stroma [11]. Toward the end of the division phase, the pyrenoid began to form in the center of the stroma, and RuBisCO is again concentrated in the pyrenoid. A 3D reconstruction of the distribution of RuBisCO (**B, left**) and a 3D reconstruction of the main pyrenoid, satellite pyrenoids, and thylakoid membranes super-imposed upon the 3D distribution of RuBisCO (**B, right**) were generated. It was found that approximately 80 % of total RuBisCO was localized to the pyrenoid region (**B**) [13]. Time courses of changes in photosynthetic CO_2-fixation and the carboxylase activity of RuBisCO in *Euglena* cells during the cell cycle in synchronized cultures are shown. From a comparison of photosynthetic CO_2-fixation with the total carboxylase activity of RuBisCO extracted from *Euglena* cells in the growth phase, it is suggested that carboxylase in the pyrenoid functions in CO_2-fixation during photosynthesis [11]. Gold dots: RuBisCO proteins. Red shows the main and smaller satellite pyrenoids. Thylakoids (green). Scale bars: 5 μm. This figure is adapted from [11].

References

11. Osafune T, Yokota A, Sumida S, Hase E (1990) Immunogold localization of ribulose-1, 5-bisphosphate carboxylase/oxygenase with reference to pyrenoid morphology in chloroplasts of synchronized *Euglena gracilis* cells. Plant Physiol 92:802–808
12. Schwartzbach SD, Osafune T (eds) (2010) Immunoelectron microscopy: methods and protocols. Hamana Press, New York
13. Osafune T (2005) Immunogold localization of photosynthetic proteins in *Euglena*. Plant Morphol 17:1–13

Contributors

Tetsuaki Osafune[1]*, Tomoko Ehara[2], Shuji Sumida[2], [1]Beppubay Research Institute for Applied Microbiology, 5-5 Kajigahama, Kitsuki 873-0008, Japan, [2]Department of Microbiology, Tokyo Medical University, Shinjuku, Tokyo 160-8402, Japan
*E-mail: osafunet@aol.com

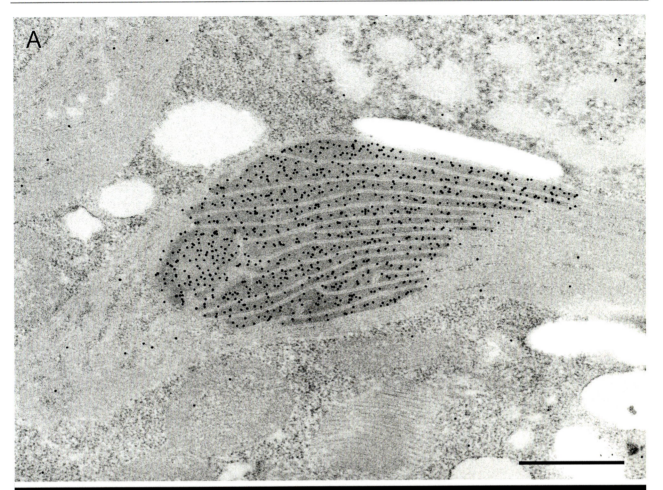

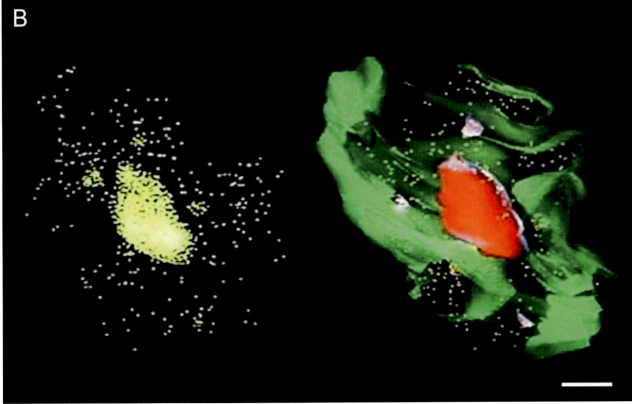

Plate 3.5

Developing and degenerating chloroplasts in *Haematococcus pluvialis*

Haematococcus pluvialis is a freshwater green algal species and is well known for its accumulation of the strong antioxidant astaxanthin, which is used in aquaculture, various pharmaceuticals, and cosmetics. High levels of astaxanthin are accumulated in cysts. It is not understood, however, how high levels of astaxanthin, which is soluble in oil, accumulate during possible during encystment. Ultrastructural 3D reconstruction was performed based on over 350 serial sections per cell to visualize the dynamics of astaxanthin accumulation and subcellular changes during the encystment of *H. pluvialis*. This study showcases marked dynamics of subcellular elements, such as chloroplast degeneration, in the transition from green coccoid cells to red cyst cells during encystment. In green coccoid cells, chloroplasts account for 41.7 % of the total cell volume and the relative volume of astaxanthin was very low (0.2 %). In contrast, oil droplets containing astaxanthin predominate in cyst cells (52.2 %), in which the total chloroplast volume is markedly decreased (9.7 %). Volumetric measurements also demonstrate that the relative volumes of the cell wall, starch grains, pyrenoids, mitochondria, Golgi apparatus, and the nucleus are smaller in a cyst cell than in green coccid cells. This indicates that chloroplasts are degraded, resulting in a net-like morphology, but chloroplasts do not completely disappear in the red cyst stage [14].

This figure shows 3D images of a green coccid cell (**A**) and a red cyst cell (**B**). Cells of *H. pluvialis* were fixed with 2.5 % glutaraldehyde and 2.5 % $KMnO_4$ (green coccid cells) or 2.5 % glutaraldehyde and 1 % OsO_4 (red cyst cells). Ultrathin serial-sections were cut on a ultramicrotome using a diamond knife, and images were obtained using a transmission electron microscope (TEM) at 100 kV. Digital TEM images were trimmed using Adobe Photoshop and printed on A4 paper sheets. Contours of each subcellular element were traced manually using color marker pens, and then the images were scanned and converted into digital images (JPG format). 3D images were subsequently reconstructed using software. Scale bar: 5 μm. This figure is adapted from [14].

Contributors

Shuhei Ota, Shigeyuki Kawano*, Department of Integrated Biosciences, Graduate School of Frontier Sciences, The University of Tokyo, Bldg. FSB-601, 5-1-5 Kashiwanoha, Kashiwa, Chiba 277-8562, Japan
*E-mail: kawano@k.u-tokyo.ac.jp

References

14. Wayama M, Ota S, Matsuura H, Nango N, Hirata A, Kawano S (2013) Three-dimensional ultrastructural study of oil and astaxanthin accumulation during encystment in the green alga *Haematococcus pluvialis*. PLoS One 8:e53618

3 Chloroplasts

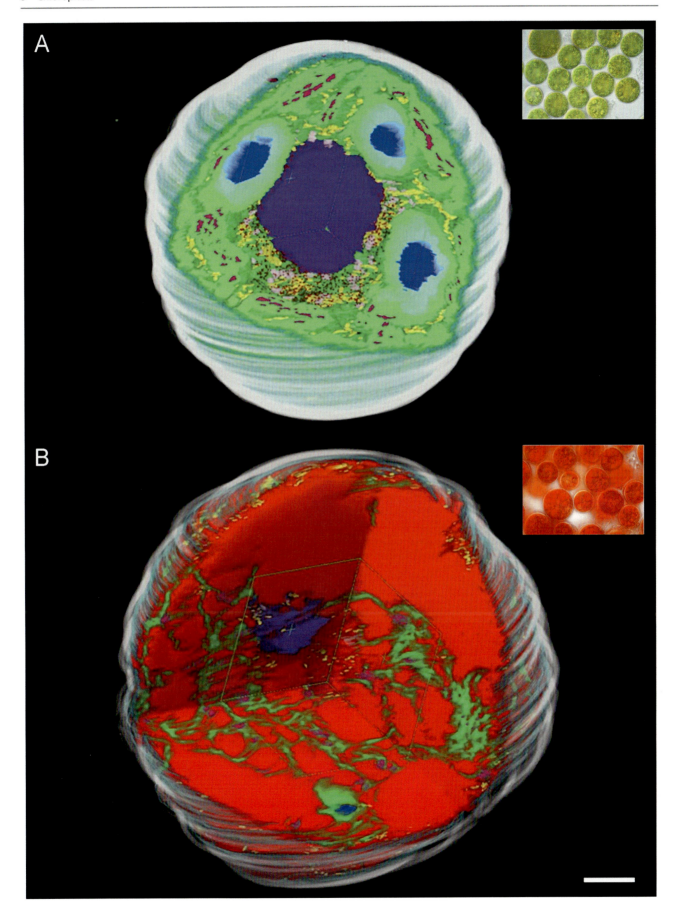

Plate 3.6

Monoplastidic cells in lower land plants

Although the cells of seed plants contain a few dozen to hundreds of plastids, many lower land plants (archegoniate plants) have cells that contain only a single plastid [15, 16]. The cells in vegetative dividing tissues of hornworts and some microphyllophytes (*Isoetes* and *Selaginella*) are typically monoplastidic (contain only one or two chloroplasts). In reproductive cell lineages, monoplastidic conditions are more commonly observed. For example, sporocytes (spore mother cells) of bryophytes (all mosses, hornworts, and some liverworts), microphyllophytes and some ferns (*Isoetes*, *Selaginella*, *Lycopodium* and *Angiopteris*) are typically monoplastidic. Additionally, sperm cells of all bryophytes and some microphyllophytes are monoplastidic. Since DNA containing organelles (nucleus, mitochondria and plastid) do not arise de novo in the cell, they should be transmitted to each daughter cell through cell division. In monoplastidic cells, morphogenetic plastid division seems to ensure the allotment of a plastid to the daughter cells. The single plastid divides before nuclear division and divided plastids serve as microtubule organizing centers for mitotic apparatuses such as the centrosome [17] (see Plate 6.5). The monoplastidic state of green algae and many archegoniate plants is believed to be an ancestral character. During the evolution of land plants, establishment of the polyplastidic condition might have allowed a more effective plastid distribution for photosynthesis.

The upper figure (**A**) shows monoplastidic cells in the gametophytic vegetative tissue of a hornwort, *Anthoceros punctatus*. Each chloroplast contains a pyrenoid similar to that of green algae (see Plate 3.4). The lower figure (**B**) shows a sporocyte of a liverwort, *Blasia pusilla*. In meiotic prophase, the single chloroplast divides twice in advance of meiotic nuclear division and the four resultant chloroplasts migrate to equidistant positions in the sporocyte. The sporocyte cytoplasm then lobes into four future spore domains associated with chloroplast division and migration. Scale bars: 10 μm.

References

15. Shimamura M, Itouga M, Tsubota H (2012) Evolution of apolar sporocytes in marchantialean liverworts: implications from molecular phylogeny. J Plant Res 125:197–206
16. Shimamura M, Mineyuki Y, Deguchi H (2003) A review of the occurrence of monoplastidic meiosis in liverworts. J Hattori Bot Lab 94:179–186
17. Shimamura M, Brown RC, Lemmon BE, Akashi T, Mizuno K, Nishihara N, Tomizawa KI, Yoshimoto K, Deguchi H, Hosoya H, Horio T, Mineyuki Y (2004) γ-Tubulin in basal land plants: characterization, localization, and implication in the evolution of acentriolar microtubule organizing centers. Plant Cell 16:45–59

Contributors

Masaki Shimamura*, Department of Biological Science, Faculty of Science, Hiroshima University, 1-3-1 Kagamiyama, Higashi-Hiroshima, Hiroshima 739-8526, Japan
*E-mail: mshima@hiroshima-u.ac.jp

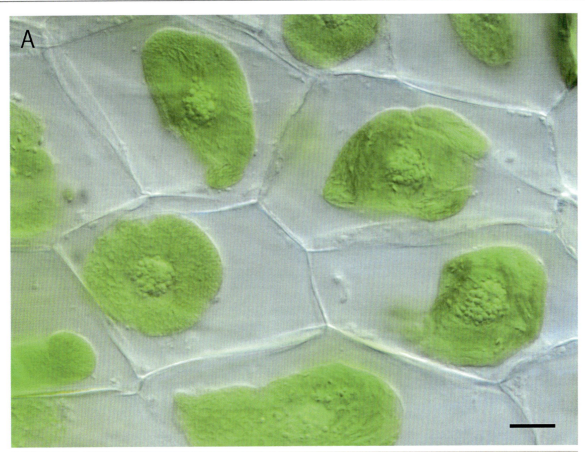

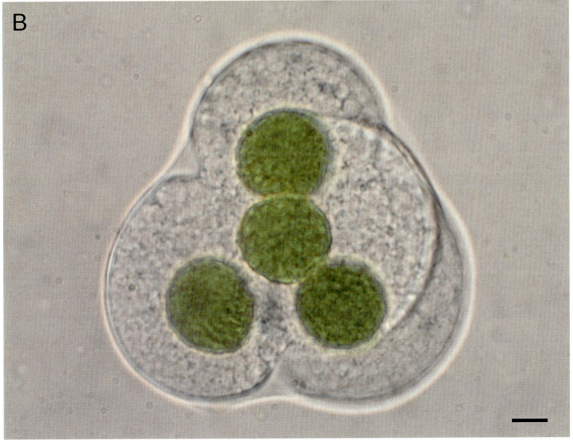

Plate 3.7

Dimorphic chloroplasts in the epidermis of the aquatic angiosperm *Podostemaceae* family

Plants of the *Podostemoideae*, a subfamily of the unique aquatic angiosperm family *Podostemaceae*, which are found in rapids and waterfalls of the tropics and subtropics, have two different sized chloroplasts in each epidermal cell [18]. Such dimorphic chloroplasts have not been reported in any other angiosperm, suggesting that the dimorphism represents an adaptation to the unique habitat of the *Podostemaceae*. Plant bodies in this family are submerged in water and subjected to torrential water during the rainy season, and emerge above the water surface to produce flowers and fruits in the dry season.

This figure shows the small and the large chloroplasts in the root epidermal cells of *Hydrobryum khaoyaiense*. Small and large chloroplasts are located separately along the upper and lower tangential walls of each epidermal cell, respectively, with no chloroplasts along the radial walls (**A**). Small chloroplasts are approximately one-sixth the size of large chloroplasts, significantly smaller in area (3.7 ± 1.8 μm^2, N = 70 chloroplasts from 15 cells) than the large ones (23.7 ± 11.0 μm^2, N = 37 from 15 cells) (Student's t-test, p < 0.01) [18]. Magnified images of one small chloroplast and one large chloroplast are shown in **B** and **C**, respectively. Ultrastructurally, small and large chloroplasts both contain normal grana and osmiophilic granules. Large chloroplasts haver four to five (and sometimes up to eight) thylakoid layers per granum, and well developed large starch grains (**C**). Large chloroplasts are identical to chloroplasts in mesophyll and parenchyma cells in their size and ultrastructure. On the other hand, small chloroplasts contained three to four thylakoid layers per granum, but have very few starch grains (**B**), and hence, the small chloroplasts may perform unique functions. *Podostemaceae* utilize HCO_3^- as a source of CO_2 for photosynthesis like the majority of submerged freshwater angiosperms. Therefore, it may be possible that the small chloroplasts function mainly to supply energy to CO_2 uptake process via HCO_3^- pump.

Roots of *Hydrobryum khaoyaiense* M. Kato were fixed in the field with 1.6 % glutaraldehyde (GA) in river water (Haew Narok Waterfall, Khao Yai National Park, Thailand) to avoid chloroplast movement in response to changing light directions. All fixed materials were kept at 4 °C for 24 h, then post-fixed in 1.0 % osmium tetroxide in 0.05 M phosphate buffer (pH 7.2) for 1 h at 4 °C. Samples were dehydrated in an ethanol series and embedded in epoxy resin (Plain Resin; Nissin EM). For TEM, ultrathin sections (70 nm thick) were stained with uranyl acetate and lead citrate and imaged using a transmission electron microscope. Ch^{-L}, large chloroplast; Ch^{-S}, small chloroplast. Scale bars: 5 μm (**A**), 1 μm (**B, C**).

Contributors

Rieko Fujinami*, Department of Chemical and Biological Sciences, Japan Women's University, 2-8-1 Mejirodai, Tokyo 112-8681, Japan
*E-mail: fujinamir@fc.jwu.ac.jp

References

18. Fujinami R, Yoshihama I, Imaichi R (2011) Dimorphic chloroplasts in the epidermis of Podostemoideae, a subfamily of the unique aquatic angiosperm family Podostemaceae. J Plant Res 124:601–605

3 Chloroplasts

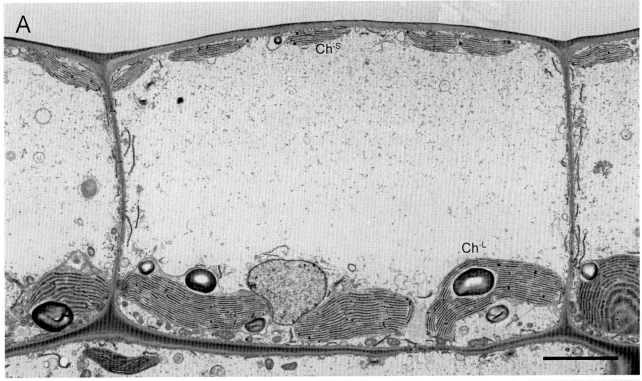

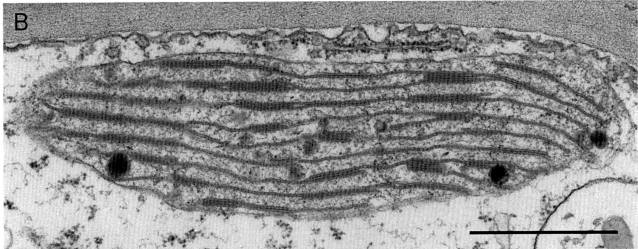

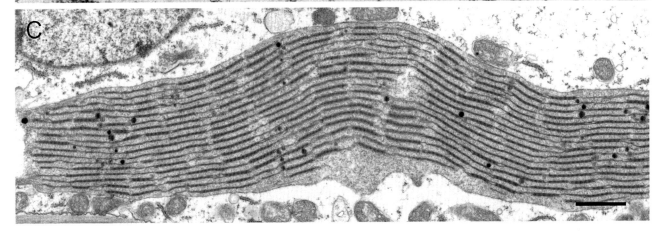

Plate 3.8

Distribution of chloroplasts and mitochondria in *Kalanchoë blossfeldiana* mesophyll cells

In leaves of some succulent crassulacean-acid-metabolism (CAM) plants, including species of the *Kalanchoë* genus, it was found that chloroplasts clumping is induced by a combination of light and water stress. In leaves of plants withheld from water for 10 days (water-stressed plant), chloroplasts clump densely in under light conditions and disperse during darkness. Chloroplast clumping results in leaf optical changes, namely a decrease in absorbance and an increase in transmittance. The plant stress hormone abscisic acid induces chloroplast clumping in leaf cells under light conditions, suggesting that this phenomenon in succulent plants is a morphological mechanism that protects against light stress intensified by a severe water deficiency [19].

Shown here are confocal microscopic images showing the distribution of chloroplasts (red, autofluorescence) and mitochondria (green, Rhodamine 123) in mesophyll cells of well-watered *Kalanchoë blossfeldiana* plants (**A**) and water-stressed plants (**B**) [20]. The inset figure is an enlargement of the rectangle in (**A**). In the leaves of well-watered plants, chloroplasts and mitochondria in mesophyll cells are dispersed across a wide area (**A**), and some mitochondria are located adjacent to each chloroplast (**inset** in **A**). In the leaves of water-stressed plants exposed to light, chloroplasts become densely clumped in mesophyll cells (**B**). The intracellular locations of the chloroplast clumps varied. Although many mitochondria were observed in the chloroplast clump, they were also distributed to other areas of the mesophyll cells (**B**). There were a few mitochondria around each chloroplast under both well-watered and water-stressed conditions. The positioning of mitochondria adjacent to chloroplasts seems to be essential for chloroplast functioning [20].

Leaf segments were hand-sectioned with a razor blade and stained with fluorescent agents under dark conditions. For mitochondria, sections were incubated in 1 µg/mL Rhodamine 123 solution for 2 m; leaf segments were not fixed because Rhodamine 123 cannot stain mitochondria under such conditions. Stained sections were observed under a confocal laser microscope. Clumped chloroplasts, cCH; mesophyll cell, Me; chloroplast, arrow; mitochondria, arrowhead. Scale bars: 50 µm (**main**), 10 µm (**inset**). This figure is adapted from [20].

Contributors

Ayumu Kondo*, Faculty of Agriculture, Meijo University, 1-501 Shiogamaguchi, Tempaku, Nagoya 468-8502, Japan
*E-mail: ayumu@meijo-u.ac.jp

References

19. Kondo A, Kaikawa J, Funaguma T, Ueno O (2004) Clumping and dispersal of chloroplasts in succulent plants. Planta 219:500–506
20. Kondo A, Shibata K, Sakurai T, Tawata M, Funaguma T (2006) Intracellular positioning of nucleus and mitochondria with clumping of chloroplasts in the succulent CAM plant *Kalanchoë blossfeldiana*: an investigation using fluorescence microscopy. Plant Morphol 18:69–73

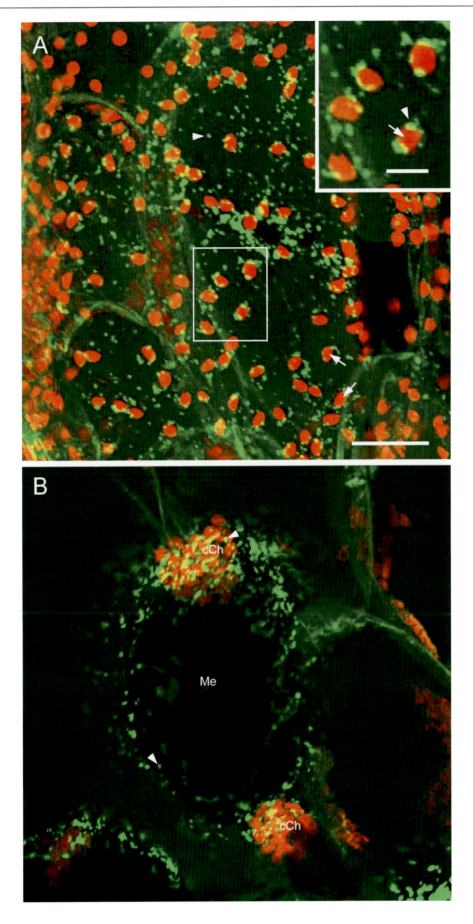

Plate 3.9

Etioplast prolamellar bodies in *Arabidopsis thaliana* etiolated cotyledon

When plants are grown in the dark cotyledons are white or yellow and plastids undergo limited development in etiolated cotyledon. Etioplasts are present in the cells of etiolated cotyledon. Upon introduction to light, normal growth resumes and the etiolated cotyledon turns green and begins to photosynthesize. The etioplast is a special state in transitioning from the proplastid to a normal, fully functioning chloroplast. Etioplasts are not an intermediate in normal chloroplast development. Proplastids become chloroplasts in plants grown in the light, or become etioplasts in plants kept in darkness. The characteristic proteins of mature chloroplasts are absent or present in very low amounts in etioplasts, although they contain some of the lipids and a precursor pigment called protochlorophyllide, but no chlorophyll [21]. The continued synthesis of lipids without synthesis of thylakoid protein leads to the structure of the prolamellar body which consists of tubes that branch in three dimensions. The prolamellar bodies form a quasi-crystalline lattice whose continuous surface is curved in opposite directions with a continuous compartment inside the tubes. The tetrahedral membrane lattice is the most common arrangement featuring branched, tubular repeating units interconnected in three dimensions.

In order to detect the membranes of prolamellar bodies, endplasmic reticulums and Golgi apparatus in the cell we performed the following fixation. Etiolated cotyledons were collected from plants grown on growth medium for 5 days in the dark at 22 °C. Samples were fixed for 3 h at 4 °C in cacodylate buffer (pH 7.4) containing 4 % paraformaldehyde, 1 % glutaraldehyde, and 0.1 M $CaCl_2$ washed with 0.1 M cacodylate buffer for 1.5 h, postfixed with 2 % OsO_4 plus 0.8 % $K_3Fe(CN)_6$ and 1 μM $CaCl_2$ in 0.1 M cacodylate buffer for 2 h at room temperature, dehydrated serially in ethanol, embedded in Spurr resin, ultrathin-sectioned, stained with uranium and lead, and observed with an electron microscope [22]. Scale bar: 500 μm.

Contributors

Yasuko Hayashi*, Department of Environmental Science, Graduate School of Science and Technology, Niigata University, Ikarashi, Niigata 950-2181, Japan
*E-mail: yhayashi@env.sc.niigata-u.ac.jp

References

21. Gunning BES, Steer MW (1996) Plastids. In: Gunning BES, Steer MW (eds) Plant cell biology, structure and function. Jones and Bartlett publishers, Canada, pp 20–29
22. Hayashi Y, Hayashi M, Hayashi H, Hara-Nishimura I, Nishimura M (2001) Direct interaction between glyoxisomes and lipid bodies in cotyledons of the *Arabidopsis thaliana* ped1 mutant. Protoplasma 218:83–94

3 Chloroplasts

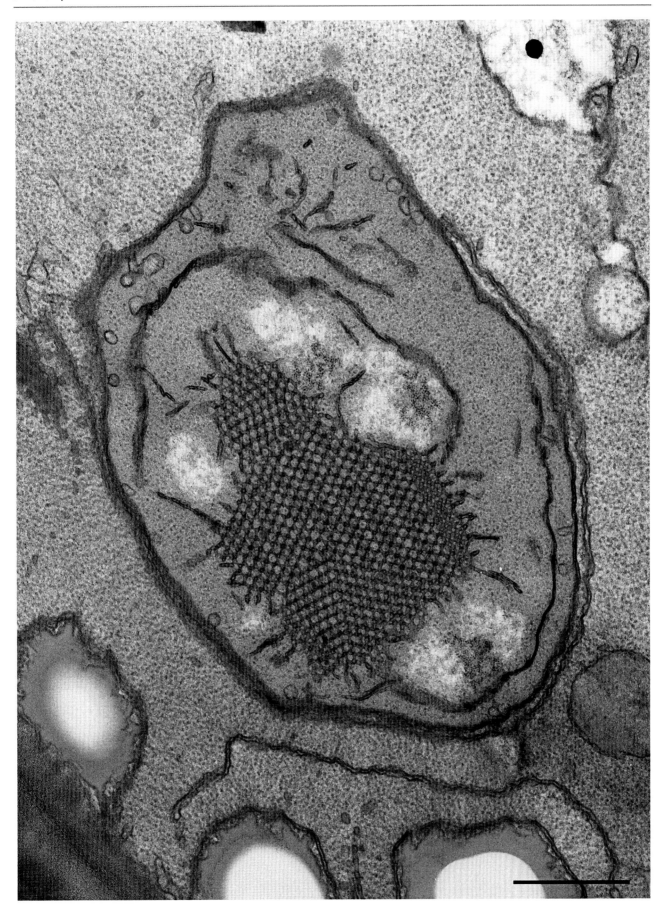

Plate 3.10

Chloroplasts and mitochondria in *Sorghum* bundle sheath cells

The vascular bundle of Sorghum leaf was observed by fluorescent microscopy (**A**) and transmission electron microscopy (**B**). Distinct structural differences between chloroplasts in mesophyll cells and bundle sheath cells are starkly apparent.

Before fixation, Sorghum leaf was cross-sectioned (<0.2 mm in thickness) by razor blade, and then cut into small pieces (<2 mm in length). Samples were then prepared for both transmission electron microscopy and fluorescent microscopy. For transmission electron microscopy, samples were rapidly frozen by high-pressure freezing. Frozen samples were then transferred to 4 % OsO_4 in anhydrous acetone and kept at $-80\,°C$ for 4 days. Samples were held at $-20\,°C$ for 2 h, $4\,°C$ for 2 h, then at room temperature for 10 min, and washed with anhydrous acetone. Samples were then embedded in Spurr resin. Ultrathin sections were cut, stained, and observed under a transmission electron microscope. For fluorescent microscopy, samples were fixed with 2 % glutaraldehyde, dehydrated by ethanol washes, and embedded in Technovit 7100 resin. 1 μm sections were cut, stained with DAPI, and observed by fluorescent microscope. Scale bars: 10 μm (**A**), 5 μm (**B**).

Contributors

Chieko Saito[1]*, Yoshihiro Kobae[2], Takashi Sazuka[3], [1]Department of Biological Sciences, Graduate School of Science, The University of Tokyo, 7-3-1 Hongo, Bunkyo-ku, Tokyo 113-0033, Japan, [2]Faculty of Agriculture, Graduate School of Agricultural and Life Sciences, The University of Tokyo, 1-1-1 Yayoi, Bunkyo-ku, Tokyo 113-8657, Japan, [3]Bioscience and Biotechnology Center, Nagoya University, Furo-cho, Chikusa-ku, Nagoya 464-8601, Japan
*E-mail: chiezo@bs.s.u-tokyo.ac.jp

3 Chloroplasts

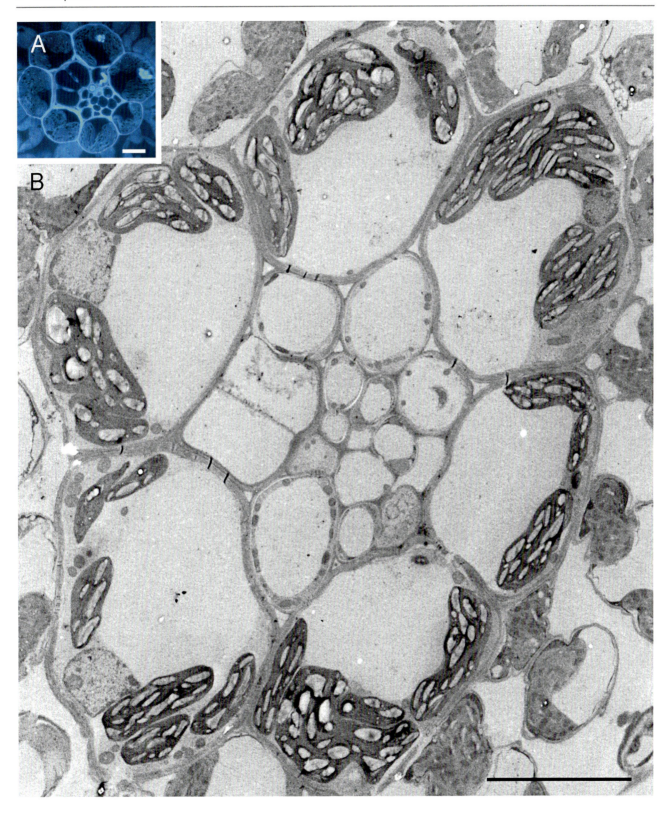

Plate 3.11

Chloroplast division machinery in *Pelargonium zonale*

Chloroplasts (plastids) ordinarily increase by binary fission, and electron-dense plastid-dividing rings (PD rings) form at the chloroplast division site. PD rings consist of inner and outer double (or triple) rings which physically divide the chloroplast. Chloroplasts are assumed to have arisen from bacterial endosymbionts, and bacterial division is instigated by the bacterial cytokinesis Z-ring protein (FtsZ). FtsZ genes have been identified in many algae and higher plants, and FtsZ localizes to co-aligned rings in chloroplasts. Immunofluorescence and electron microscopic evidence of chloroplast division via complex machinery involving the FtsZ and PD rings have been observed in the higher plant *Pelargonium zonale* Ait [23]. Prior to invagination of the chloroplast, the FtsZ protein attaches to a ring at the stromal division site. Following formation of the FtsZ ring, the inner stromal and outer cytosolic PD rings appear at the chloroplast constriction site during chloroplast division. Neither the FtsZ nor the inner PD rings change width, but the volume of the outer PD ring gradually increases. Based on these results, it appears that the FtsZ ring determines the division region, after which the inner and outer PD rings form as a lining for the FtsZ ring. With the outer ring providing the initial force, the FtsZ and inner PD rings ultimately decompose to their base components. These figures show chloroplast divisions in early embryos of *P. zonale*. Immunofluorescence images of chloroplasts labeled with anti-LlFtsZ antibody [24] are arranged from top to bottom according to chloroplast division state (chloroplasts emit red auto-fluorescence) (**A**). FtsZ (yellow-green fluorescence) was attached to rings at the division sites. Closed circular rings visualized as well as in **A** (**B**). Enlarged immunofluorescence image of dividing chloroplasts reacted with anti-LlFtsZ antibody (**C**). Electron microscopic image showing PD ring of dividing chloroplast (**D**). Insets show the chloroplast dividing site with inner and outer PD rings. Scale bars: 1 µm. This figure is adapted from [23].

Contributors

Haruko Kuroiwa*, Tsuneyoshi Kuroiwa, CREST, Initiative Research Unit, College of Science, Rikkyo University, Toshima, Tokyo 171-8501, Japan
*E-mail: haruko-k@tvr.rikkyo.ne.jp

References

23. Kuroiwa H, Mori T, Takahara M, Miyagishima S, Kuroiwa T (2002) Chloroplast division machinery as revealed by immunofluorescence and electron microscopy. Planta 215:185–190

24. Mori T, Kuroiwa H, Takahara M, Miyagishima S, Kuroiwa T (2001) Visualization of an FtsZ ring in chloroplasts of *Lilium longiflorum* leaves. Plant Cell Physiol 42:555–559

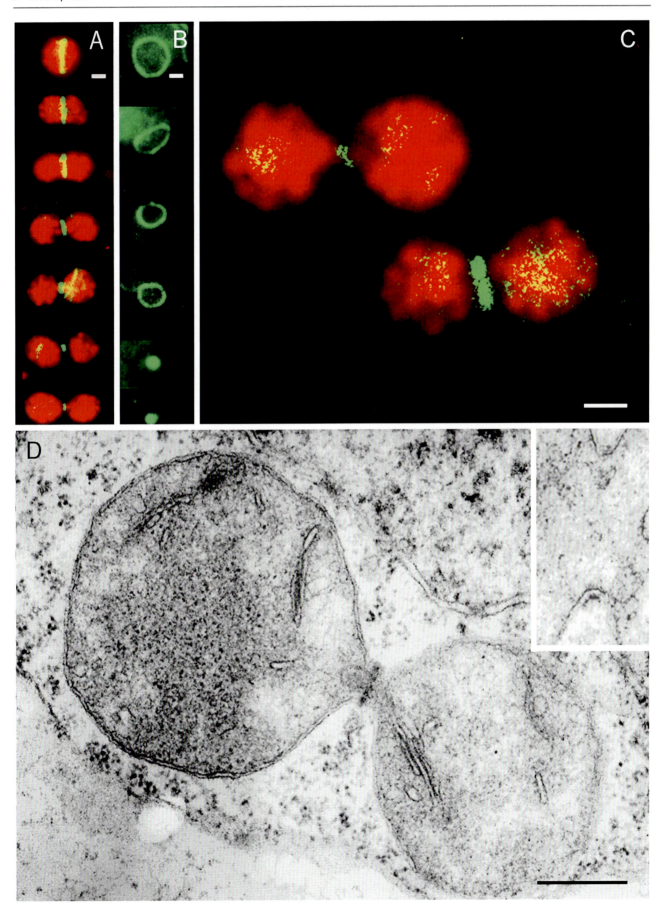

Plate 3.12

Active digestion of paternal chloroplast DNA in a young *Chlamydomonas reinhardtii* zygote

In most sexual eukaryotes, chloroplast (cp) and mitochondrial (mt) genomes are inherited almost exclusively from one parent. Uniparental inheritance of cp/mt genomes was long thought to be a passive result of the fact that eggs contain numerous organelles, while male gametes contribute, at best, only a few organelles and therefore little cp/mtDNA. However, uniparental inheritance occurs in organisms that produce gametes of identical sizes (isogamous), implying that the process is likely to be more dynamic.

In *Chlamydomonas reinhardtii*, although the maternal (mating type plus, *mt+*) and paternal (mating type minus, *mt−*) gametes are of equal sizes and contribute equal amounts of cytoplasm to the zygote, only cpDNA from the *mt+* parent is inherited due to active degradation of *mt−* cpDNA within 60 m after mating. The method by which *Chlamydomonas* selectively degrades *mt−* cpDNA has long fascinated researchers (for review [25]).

This figure represents the process of the preferential elimination of *mt−* cp nucleoids (cpDNA-protein complex: white arrow in A) in a living *C .reinhardtii* zygote. Zygotes were incubated with 1:2000 SYBR Green I solution for 5 min at room temperature. SYBR Green I staining can visualize dsDNA molecules as yellow–green fluorescence when excited by blue light. The red fluorescence shown is autofluorescence emitted from chlorophyll in chloroplasts. Cell nucleus is indicated by "N".

Young living zygotes stained by SYBR Green I (**A**). The left and right chloroplasts were derived from *mt+* and *mt−* gametes, respectively. At this time point, both of the chloroplasts have almost equal numbers of cp nucleoids. The identical zygote after 10 min (**B**). Nucleoids from the *mt−* chloroplast completely disappear, whereas mitochondrial nucleoids (arrowheads) are still visible. The active disappearance of *mt−* cp nucleoids was a rapid process that commenced about 40 m after zygote formation and was completed within 10 min. It has been confirmed that *mt−* cpDNA molecules are degraded during the disappearance of *mt−* cp nucleoids by single cell analysis using optical tweezers, indicating that the rapid disappearance of *mt−* cpDNA is the basis of uniparental inheritance [26]. A mutant defective in the active digestion of *mt−* cpDNA, *biparental (bp) 31*, was recently isolated and detailed analysis of this mutant revealed that uniparental inheritance of cpDNA is strictly controlled by the *mt+* gamete-specific homeotic gene, *GAMETE SPECIFIC PLUS (GSP) 1* [27]. Scale bar: 5 µm. This figure is adapted from [25].

Contributors

Yoshiki Nishimura*, Department of Botany, Kyoto University, Oiwake-cho, Kita-Shirakawa, Kyoto 606-8502, Japan
*E-mail: yoshiki@pmg.bot.kyoto-u.ac.jp

References

25. Nishimura Y (2010) Uniparental inheritance of cpDNA and the genetic control of sexual differentiation in *Chlamydomonas reinhardtii*. J Plant Res 123:149–162
26. Nishimura Y, Misumi O, Matsunaga S, Higashiyama T, Yokota A, Kuroiwa T (1999) The active digestion of uniparental chloroplast DNA in a single zygote of *Chlamydomonas reinhardtii* is revealed by using the optical tweezer. Proc Natl Acad Sci U S A 96:12577–12582
27. Nishimura Y, Shikanai Y, Nakamura S, Kawai-Yamada M, Uchimiya H (2012) The Gsp1 triggers sexual developmental program including inheritance of cpDNA and mtDNA in *Chlamydomonas reinhardtii*. Plant Cell 24:2401–2414

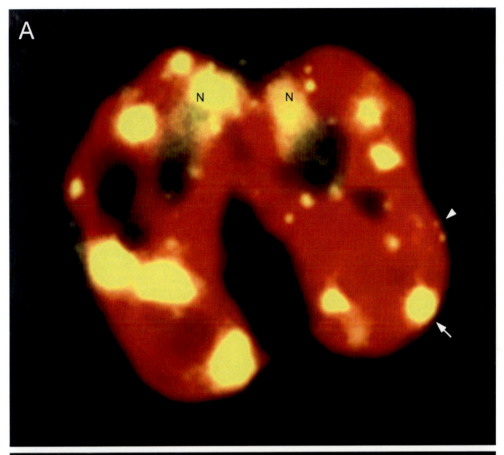
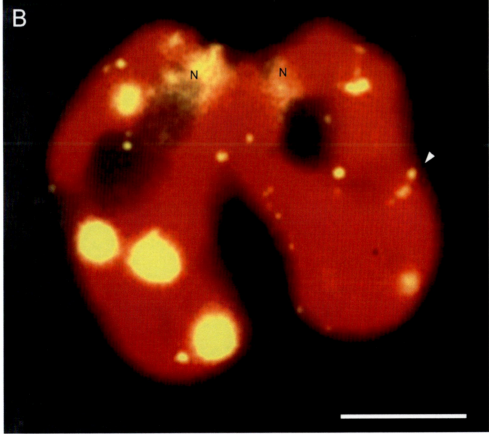

Chapter References

1. Miyagishima S, Kabeya Y (2010) Chloroplast division: squeezing the photosynthetic captive. Curr Opin Microbiol 13:738–746
2. Mita T, Kanbe T, Tanaka K, Kuroiwa T (1986) A ring structure around the dividing plane of the *Cyanidium caldarium* chloroplast. Protoplasma 130:211–213
3. Miyagishima S, Nishida K, Mori T, Matsuzaki M, Higashiyama T, Kuroiwa H, Kuroiwa T (2003) A plant-specific dynamin-related protein forms a ring at the chloroplast division site. Plant Cell 15:655–665
4. Miyagishima S, Itoh R, Aita S, Kuroiwa H, Kuroiwa T (1999) Isolation of dividing chloroplasts with intact plastid-dividing rings from a synchronous culture of the unicellular red alga *Cyanidioschyzon merolae*. Planta 209:371–375
5. Yoshida Y, Kuroiwa H, Misumi O, Yoshida M, Ohnuma M, Fujiwara T, Yagisawa F, Hirooka S, Imoto Y, Matsushita K, Kawano S, Kuroiwa T (2010) Chloroplasts divide by contraction of a bundle of nanofilaments consisting of polyglucan. Science 329:949–953
6. Yoshida Y, Fujiwara T, Imoto Y, Yoshida M, Ohnuma M, Hirooka S, Misumi O, Kuroiwa H, Kato S, Matsunaga S, Kuroiwa T (2013) The kinesin-like protein TOP promotes Aurora localization and induces mitochondrial, chloroplast and nuclear division. J Cell Sci 126:2392–2400
7. Yoshida Y, Kuroiwa H, Misumi O, Nishida K, Yagisawa F, Fujiwara T, Nanamiya H, Kawamura F, Kuroiwa T (2006) Isolated chloroplast division machinery can actively constrict after stretching. Science 313:1435–1438
8. Iino M, Hashimoto H (2003) Intermediate features of cyanelle division of *Cyanophora paradoxa* (Glaucocystophyta) between cyanobacterial and plastid division. J Phycol 39:561–569
9. Sato M, Mogi Y, Nishikawa T, Miyamura S, Nagumo T, Kawano S (2009) The dynamic surface of dividing cyanelles and ultrastructure of the region directly below the surface in *Cyanophora paradoxa*. Planta 229:781–791
10. Sato M, Nishikawa T, Kajitani H, Kawano S (2007) Conserved relationship between FtsZ and peptidoglycan in the cyanelles of *Cyanophora paradoxa* similar to that in bacterial cell division. Planta 227:177–187
11. Osafune T, Yokota A, Sumida S, Hase E (1990) Immunogold localization of ribulose-1,5-bisphosphate carboxylase/oxygenase with reference to pyrenoid morphology in chloroplasts of synchronized *Euglena gracilis* cells. Plant Physiol 92:802–808
12. Schwartzbach SD, Osafune T (eds) (2010) Immunoelectron microscopy: methods and protocols. Hamana Press, New York
13. Osafune T (2005) Immunogold localization of photosynthetic proteins in *Euglena*. Plant Morphol 17:1–13
14. Wayama M, Ota S, Matsuura H, Nango N, Hirata A, Kawano S (2013) Three-dimensional ultrastructural study of oil and astaxanthin accumulation during encystment in the green alga *Haematococcus pluvialis*. PLoS One 8:e53618
15. Shimamura M, Itouga M, Tsubota H (2012) Evolution of apolar sporocytes in marchantialean liverworts: implications from molecular phylogeny. J Plant Res 125:197–206
16. Shimamura M, Mineyuki Y, Deguchi H (2003) A review of the occurrence of monoplastidic meiosis in liverworts. J Hattori Bot Lab 94:179–186
17. Shimamura M, Brown RC, Lemmon BE, Akashi T, Mizuno K, Nishihara N, Tomizawa KI, Yoshimoto K, Deguchi H, Hosoya H, Horio T, Mineyuki Y (2004) γ-Tubulin in basal land plants: characterization, localization, and implication in the evolution of acentriolar microtubule organizing centers. Plant Cell 16:45–59
18. Fujinami R, Yoshihama I, Imaichi R (2011) Dimorphic chloroplasts in the epidermis of Podostemoideae, a subfamily of the unique aquatic angiosperm family Podostemaceae. J Plant Res 124:601–605
19. Kondo A, Kaikawa J, Funaguma T, Ueno O (2004) Clumping and dispersal of chloroplasts in succulent plants. Planta 219:500–506
20. Kondo A, Shibata K, Sakurai T, Tawata M, Funaguma T (2006) Intracellular positioning of nucleus and mitochondria with clumping of chloroplasts in the succulent CAM plant *Kalanchoë blossfeldiana*: an investigation using fluorescence microscopy. Plant Morphol 18:69–73
21. Gunning BES, Steer MW (1996) Plastids. In: Gunning BES, Steer MW (eds) Plant cell biology, structure and function. Jones and Bartlett publishers, Canada, pp 20–29
22. Hayashi Y, Hayashi M, Hayashi H, Hara-Nishimura I, Nishimura M (2001) Direct interaction between glyoxisomes and lipid bodies in cotyledons of the *Arabidopsis thaliana* ped1 mutant. Protoplasma 218:83–94
23. Kuroiwa H, Mori T, Takahara M, Miyagishima S, Kuroiwa T (2002) Chloroplast division machinery as revealed by immunofluorescence and electron microscopy. Planta 215:185–190
24. Mori T, Kuroiwa H, Takahara M, Miyagishima S, Kuroiwa T (2001) Visualization of an FtsZ ring in chloroplasts of *Lilium longiflorum* leaves. Plant Cell Physiol 42:555–559
25. Nishimura Y (2010) Uniparental inheritance of cpDNA and the genetic control of sexual differentiation in *Chlamydomonas reinhardtii*. J Plant Res 123:149–162
26. Nishimura Y, Misumi O, Matsunaga S, Higashiyama T, Yokota A, Kuroiwa T (1999) The active digestion of uniparental chloroplast DNA in a single zygote of *Chlamydomonas reinhardtii* is revealed by using the optical tweezer. Proc Natl Acad Sci USA 96:12577–12582
27. Nishimura Y, Shikanai Y, Nakamura S, Kawai-Yamada M, Uchimiya H (2012) The Gsp1 triggers sexual developmental program including inheritance of cpDNA and mtDNA in *Chlamydomonas reinhardtii*. Plant Cell 24:2401–2414

The Endoplasmic Reticulum, Golgi Apparatuses, and Endocytic Organelles

4

Tetsuko Noguchi, Sachihiro Matsunaga, and Yasuko Hayashi

Vesicular transport system is a dynamic system responsible for protein, sugar, and lipid transport. Exocytic vesicular transport begins at the endoplasmic reticulum (ER) (proteins and lipids) or the Golgi apparatus (polysaccharides) and terminates at vacuolar compartments or the cell surface. In addition to the exocytic pathway, the endocytic pathway, in which cargo is transported from the plasma membrane to lytic/vacuolar compartments or the *trans*-Golgi networks via endosomes, is also a vesicular transport system.

In this chapter, H. Ueda et al. show a remarkable three-dimensional reconstruction of the ER in living cells. The basket-like ER-network is shown in a tobacco BY-2 cell, and its rearrangement is captured in *Arabidopsis thaliana* by use of a confocal laser scanning microscope. T. Noguchi shows the typical ultrastructure of the ER with polyribosomes in the green alga *Botryococcus braunii* using rapid freezing electron microscopy. Y. Hayashi and T. Sakurai show a unique ER-derived compartment, the ER body, in *Arabidopsis thaliana*. ER bodies are captured in living cells and their ultrastructure is also visualized by high-pressure freezing electron microscopy.

The major functions of the Golgi apparatus are protein glycosylation and polysaccharide synthesis. The Golgi apparatus consists of a stack of flattened membrane–bound cisternae. As opposed to other organelles, the Golgi stack is a polar structure, with different functions across the structure; the *cis*-face is located near the ER and the *trans*-face produces secretory vesicles. The *trans*-Golgi network, with its attached clathrin coated vesicles, is the sorting site of glycoproteins carried from Golgi cisternae.

The basic molecular mechanisms of vesicular transport and the basic construction of cisternal stacks are similar between animal and plant/algal cells. On the other hand, the architecture, distribution, and reproduction of the Golgi apparatus in plant cells are quite different from these features in mammalian cells. In mammalian cells, most cisternal stacks link together and form a single large complex near the cell nucleus. In contrast, the Golgi apparatus in plant and algal cells is composed of small, separated cisternal stacks (Golgi bodies). In plant cells, 25 or fewer to several hundreds of Golgi bodies 0.5–1.5 μm in diameter are distributed throughout the cytoplasm. Because of their small size and distribution pattern in cells, the plant Golgi apparatus was confirmed about 60 years after the discovery of the Golgi apparatus in neurons by Camillo Golgi (1898). While the animal Golgi apparatus reproduces by disassembling at mitosis, many Golgi bodies in plant cells divide synchronously into two before mitosis and do not fragment during mitosis. These differences may reflect on the basic functions of the Golgi apparatus between the two kingdoms: the plant Golgi apparatus is more involved in polysaccharide secretion than protein secretion, the latter of which is the major activity of the Golgi apparatus in animal cells. In addition, vesicular transport mainly depends on microtubules in animal cells, but on actin filaments in plant cells.

M. Sato shows functioning Golgi bodies with hypertrophied *trans*-cisternae in a *Brachypodium distachyon* root cap cell using high-pressure freezing electron microscopy. T. Noguchi presents Golgi bodies in *Tradescantia reflexa* pollen grain, captured by a high-pressure freezing electron microscopy. The *trans*-Golgi networks are not as developed in plant cells as in animal cells. T. Noguchi shows the developed TGN in green alga *B. braunii* obtained by applying quick-freeze deep-etch electron microscopy.

Endocytosis is an energy-consuming process by which cells adapt molecules from the extracellular environment and plasma membrane. T. Noguchi shows clathrin-coated buds on the plasma membrane, which is the first step in the endocytic pathway, and multivesicular bodies, a kind of endosome, in *B. braunii* using rapid freezing electron microscopy. M. Fujimoto and T. Ueda successfully catch the dynamic sequential movements of dynamin-related proteins and clathrin light chain in living *A. thaliana* by variable incidence angle fluorescence microscopy.

T. Noguchi (✉)
Course of Biological Sciences, Faculty of Science, Nara Women's University, Kitauoya-nishimachi, Nara 630-8506, Japan
e-mail: noguchi@cc.nara-wu.ac.jp

S. Matsunaga
Department of Applied Biological Science, Faculty of Science and Technology, Tokyo University of Science, 2641 Yamazaki, Noda, Chiba 278-8510, Japan
e-mail: sachi@rs.tus.ac.jp

Y. Hayashi
Department of Environmental Science and Technology, Graduate School of Science and Technology, Niigata University, Ikarashi, Niigata 950-2181, Japan
e-mail: yhayashi@env.sc.niigata-u.ac.jp

T. Noguchi et al. (eds.), *Atlas of Plant Cell Structure*,
DOI 10.1007/978-4-431-54941-3_4, © Springer Japan 2014

Plate 4.1

Endoplasmic reticulum throughout the cytoplasm

The endoplasmic reticulum (ER) is the organelle with the largest membrane surface area in eukaryotic cells. The ER membrane is physically continuous and extends throughout the cytoplasm. In a three-dimensional reconstitution, the plant ER looks like a basket encompassing the cell because of the thin cytoplasmic area between the plasma membrane and the large central vacuoles in a characteristic ER network architecture. The ER network is composed of interconnected membrane tubules and cisternae, and continuously rearranges with the elongation of tubules, sliding of junctions, and closures of rings. Rearrangement of the ER network depends not only on the actin-myosin system to elongate tubules [1] but also on ER membrane fusogens to interconnect elongated tubules and cisternae.

A striking feature of the plant ER is the active streaming driven by plant-specific class XI myosins [2, 3]. When the ER network, with its large surface area, moves in one direction along actin filament bundles, it pulls the cytosol along, resulting in passive cytoplasmic streaming. The ER seems to be a natural candidate to drive cytoplasmic streaming, the most well known intracellular motility in plant cells.

Micrographs show ER visualized with an ER-localized green fluorescent protein (GFP). Sequential images of a single tobacco BY-2 cell expressing ER-localized GFP were taken along the optical z-axis (0.33 μm intervals) with a laser-scanning confocal microscope (LSM780, Carl Zeiss) and projections were reconstituted from 45 sequential images (**A**). Time-lapse images of peripheral ER in a cotyledon epidermal cell in transgenic *Arabidopsis thaliana* expressing ER-localized GFP were taken at 1-s intervals with a microscope equipped with a confocal laser scanning unit (CSU X1, Yokogawa Electric) and an EM-CCD camera (iXon3, ANDOR) (**B–D**). Rearrangement of the ER network was observed: a new tubule (arrowhead) emerged, elongated, and fused to other tubules within 2 s. Asterisks, ER bodies. Scale bars: 10 μm.

Contributors

Haruko Ueda[1]*, Etsuo Yokota[2], Ikuko Hara-Nishimura[1], [1]Department of Botany, Graduate School of Science, Kyoto University, Kyoto 606-8502, Japan, [2]Department of Life Science, Graduate School of Life Science, University of Hyogo, Hyogo 678-1297, Japan
*E-mail: hueda@gr.bot.kyoto-u.ac.jp

References

1. Yokota E, Ueda H, Hashimoto K, Orii H, Shimada T, Hara-Nishimura I, and Shimmen T (2011) Myosin XI-dependent formation of tubular structures from endoplasmic reticulum isolated from tobacco cultured BY-2 cells. Plant Physiol 156:129–143
2. Ueda H, Yokota E, Kutsuna N, Shimada T, Tamura K, Shimmen T, Hasezawa S, Dolja VV, Hara-Nishimura I (2010) Myosin-dependent endoplasmic reticulum motility and F-actin organization in plant cells. Proc Natl Acad Sci U S A 107: 6894–6899
3. Yokota E, Ueda S, Tamura K, Orii H, Uchi S, Sonobe S, Hara-Nishimura I, Shimmen T (2009) An isoform of myosin XI is responsible for the translocation of endoplasmic reticulum in tobacco cultured BY-2 cells. J Exp Bot 60:197–212

4 The Endoplasmic Reticulum, Golgi Apparatuses, and Endocytic Organelles

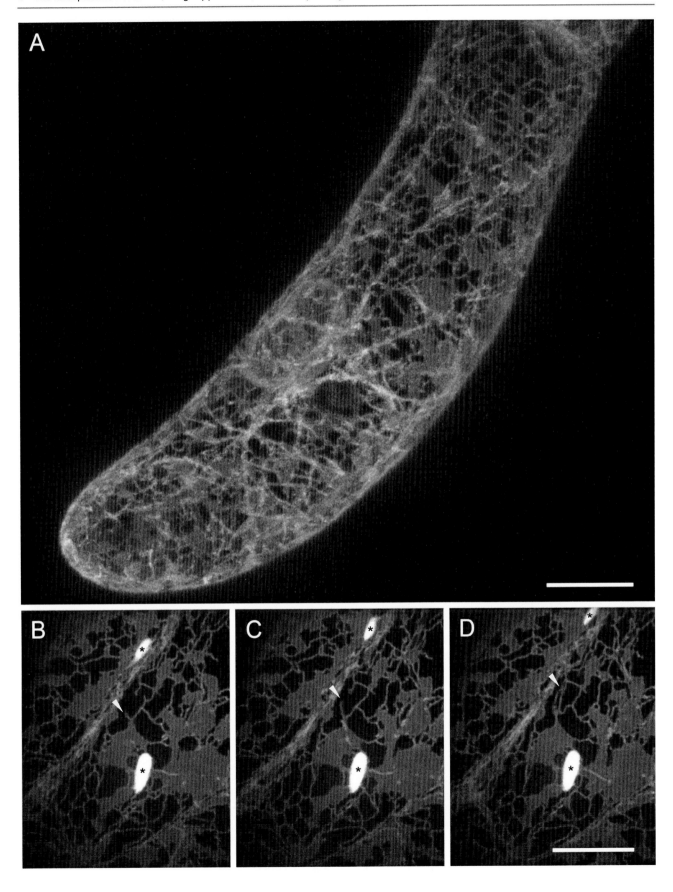

Plate 4.2

Endoplasmic reticulum in the green alga *Botryococcus braunii*

The endoplasmic reticulum (ER) is a network of flattened sacs and tubules that serve a variety of functions in the cell. It consists of two regions; the rough ER (rER), attached with ribosomes on the surface (**A**), and the smooth ER (sER), which is without ribosomes. On the rER cut parallel to its surface (ER*), polyribosomes (**B**) composed of several ribosomes that simiultaneously translate the same mRNA molecule, are prominent.

The green alga *Botryococcus braunii* accumulates large amounts of hydrocarbons in the extracellular space (white * in (**A**), see Plate 5.7). The initial step of the hydrocarbon synthesis pathway occurs in chloroplasts after which precursors are transferred to the plasma membrane via rER. The cortical rER near the plasma membrane is prominent throughout the cell. In addition, the rER is often in contact with a chloroplast.

Ribosomes are lost from rER budding sites to form transport vesicles containing cargo proteins which are transported to the Golgi apparatus. This transitional ER is a boundary for quality control in protein secretion [4]. In *B. braunii*, another type transitional ER region was discovered in the cortical ER and the ER at the surface of chloroplasts. These regions never bud but closely contact the plasma membrane or the chloroplast, and lose ribosomes from the surface facing the plasma membrane or the chloroplast (arrow in (**A**)), respectively [5, 6].

This figure shows transmission electron micrographs of *Botryococcus braunii* race A. Cells were attached to formvar films mounted on copper wire loops 8 mm in diameter and frozen in liquid propane at $-190\,°C$. They were then transferred to acetone $(-85\,°C)$ containing 2 % osmium tetroxide and 0.2 % uranyl acetate. After 48 h at $-85\,°C$, samples were gradually warmed to room temperature, washed with acetone, and embedded in Spurr resin, and imaged by a transmission electron microscope at 80 kV. Ch, chloroplast; ER, endoplasmic reticulum cut perpendicularly to surface; ER*, endoplasmic reticulum cut parallel to the surface; PM, plasma membrane. Scale bar: 500 nm (**A**), 50 nm (**B**).

Contributors

Tetsuko Noguchi*, Course of Biological Sciences, Faculty of Science, Nara Women's University, Kitauoya-nishimachi, Nara 630-8506, Japan
*E-mail: noguchi@cc.nara-wu.ac.jp

References

4. Farquhar MG, Hauri H-P (1997) Protein sorting and vesicular traffic in the Golgi apparatus. In: Berger EG, Roth J (eds) The Golgi apparatus. Birkhaüser Verlag, Basel/Boston/Berlin, pp 63–129
5. Hirose M, Mukaida F, Okada S, Noguchi T (2013) Active hydrocarbon biosynthesis and accumulation in a green alga, *Botryococcus braunii* (Race A) Eukaryot Cell 12(8):1132–1141
6. Suzuki R, Ito N, Uno Y, Nishii I, Kagiwada S, Okada S, Noguchi T (2013) Transformation of lipid bodies related to hydrocarbon accumulation in a green alga, *Botryococcus braunii* (Race B). PLoS One 8(12):e81626. doi:10.1371/journal.pone.0081626

4 The Endoplasmic Reticulum, Golgi Apparatuses, and Endocytic Organelles

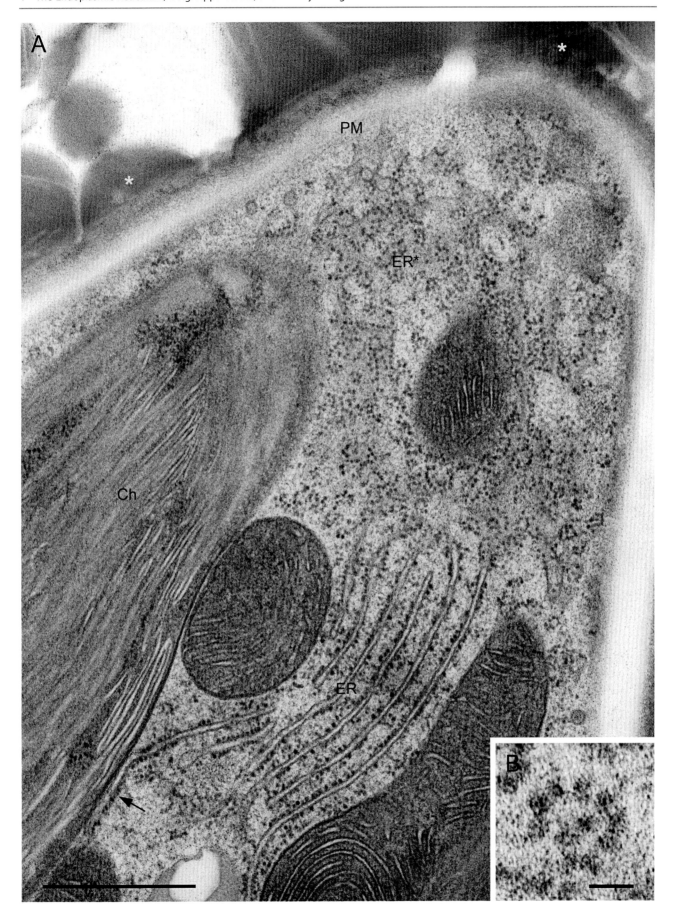

Plate 4.3

ER body in cotyledon epidermal cells

The endoplasmic reticulum (ER) is a multifunctional organelle whose most prominent functions include the synthesis of membrane lipids and secretory proteins. Several ER-derived structures with specific functions develop in plant cells. A novel ER-derived compartment in *Arabidopsis*, the ER body, was reported in 2001 [7]. This structure has the same electron-density as peroxisomes in chemically fixed ultrathin sections examined by electron microscopy but appears to have a different electron-density from peroxisomes in samples prepared by high-pressure freezing and freeze substitution methods. Shown here are spindle-shaped structures that are surrounded by membranes with ribosomes (main panel). In the 1960s, similar structures were found in some *Brassicaceae* plants. However, their main components and biological functions were not clarified at the time. In transgenic *Arabidopsis* expressing green fluorescent protein (GFP) with an ER-retention signal (KDEL), the ER network and ER bodies can be visualized (**inset**). Cotyledonary epidermal cells develop many ER bodies which decrease in number during senescence. Hypocotyls and root cells also have ER bodies, but in these tissues ER body numbers do not decrease during senescence. In contrast, rosette leaf cells of mature plants have no ER bodies. However, wound stress in rosette leaves induced formation of spindle-shaped structures similar to ER bodies. Treatment with methyl jasmonate (MeJA) in rosette leaves also induced spindle-shaped structures [8]. A nai1 mutant that does not develop ER bodies in whole seedlings was isolated in 2003. A β-glucosidase called PYK10 was identified as the main component of ER bodies by comparative analysis between the nai1 mutant and the wild type [9]. The formation of ER bodies in rosette leaves following wound stress and the putative biological function of PYK10 suggest that the ER body has a role in defense against herbivores.

For the fluorescence micrograph, *Arabidopsis thaliana* was transformed with a chimeric gene consisting of pumpkin 2S albumin signal peptide followed by GFP and a 12-amino-acid sequence including the ER-retention signal, HDEL. Specimens were examined with a laser-scanning confocal microscope equipped with an argon laser and a fluorescein filter set (excitation 465–505 nm, emission 505–550 nm). For the electron micrograph, 5-day-old cotyledons were frozen with a high-pressure freezing machine (HPM-100, Balzers), and then dehydrated for 2 days at -85 °C in acetone containing 2 % (w/v) osmium tetraoxide, dehydrated serially in ethanol, embedded in Spurr resin, ultrathin-sectioned, stained with uranium and lead, and observed with an electron microscope. Scale bars: 5 μm (**inset**), 0.5 μm (**main**).

Contributors

Yasuko Hayashi*, Toshiyuki Sakurai, Department of Environmental Science, Graduate School of Science and Technology, Niigata University, Ikarashi, Niigata 950-2181, Japan
*E-mail: yhayashi@env.sc.niigata-u.ac.jp

References

7. Hayashi Y, Yamada K, Shimada T, Matsushima R, Nishizawa NK, Nishimura M, Hara-Nishimura I (2001) A proteinase-sorting body that prepares for cell death or stresses in the epidermal cells of Arabidopsis. Plant Cell Physiol 42:894–899
8. Matsushima R, Hayashi Y, Kondo M, Shimada T, Nishimura M, Hara-Nishimura I (2002) An endoplasmic reticulum-derived structure that is induced under stress conditions in Arabidopsis. Plant Physiol 130:1807–1814
9. Matsushima R, Kondo M, Nishimura M, Hara-Nishimura I (2003) A novel ER-derived compartment, the ER body, selectively accumulates a beta-glucosidase with an ER-retention signal in Arabidopsis. Plant J 33:493–502

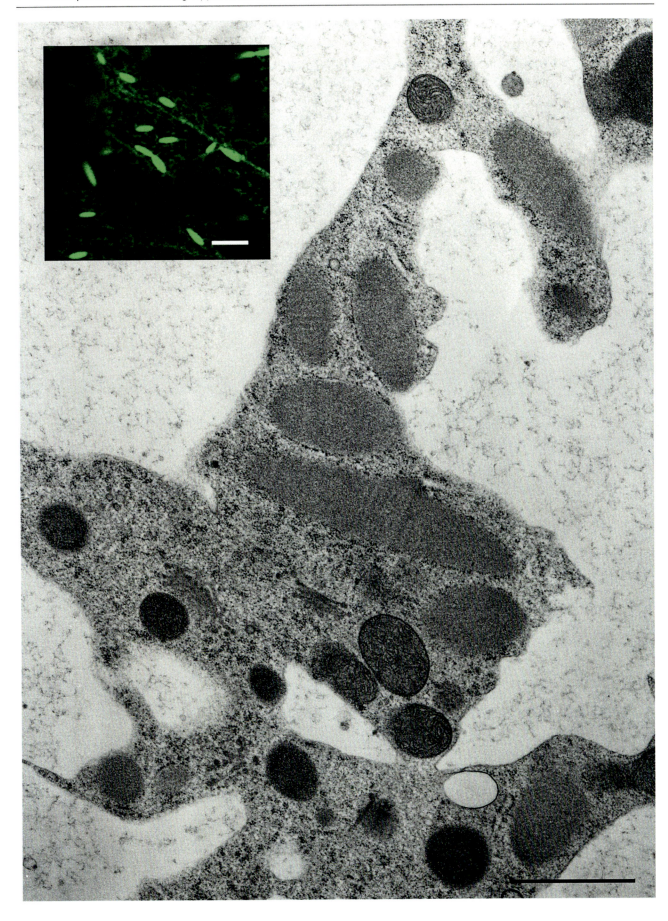

Plate 4.4

Golgi apparatuses in a *Brachypodium* root cap peripheral cell

In a plant root tip, Golgi apparatus morphology differs by cell type. For example, Golgi apparatuses in *Arabidopsis* and *Nicotiana* root cap cells change during development from meristematic cells to columella cells and to peripheral cells [10]. In a meristematic cell, Golgi apparatuses are small and compact. Golgi stacks become wide, and vesicles associated with *trans* cisternae become large, round and electron-dense as cells differentiate. Peripheral cells contain many prominent Golgi apparatuses, since the cells secrete mucilage to protect root tissue. In these Golgi apparatuses, electron-dense *trans* cisternae become hypertrophied at each end, although low density *cis* cisternae retain flat structures. Such swelling of Golgi stacks in mature peripheral cells is observed in both monocotyledons and dicotyledons [10]. At later stages of development, peripheral cells degenerate and finally slough from the root cap.

Brachypodium distachyon (Poaceae, Pooideae) is a model grass. Its root has a closed type root apical meristem and a clear root cap boundary. This electron micrograph shows a cell on the outermost layer of a root cap. The cell contains a large number of prominent Golgi apparatuses with electron-dense, hypertrophied *trans* cisternae. There are also electron-dense vesicles, which may have been secreted from these Golgi apparatuses. On the other hand, there is no such hypertrophied Golgi apparatus in the medial region of the root cap.

A root tip of *B. distachyon* was fixed by high-pressure freezing and freeze substitution and imaged by transmission electron microscopy. *B. distachyon* roots were excised with a razor blade and frozen in a high-pressure freezing machine (EM-PACT, Leica Microsystems). Samples were immediately transferred to 2 % OsO_4 in anhydrous acetone at -80 °C and incubated for 9 days. Samples were then warmed gradually from -80 to -20 °C over 20 h, from -20 to 4 °C over 24 h, and incubated for 4 h at 4 °C, and finally for 2 h at room temperature. Samples were washed three times with anhydrous acetone at room temperature and infiltrated with propylene oxide followed by increasing concentrations of Epon812 resin (TAAB) and finally embedded in the same resin. Ultrathin sections (60–70 nm) were cut with a diamond knife on an ultramicrotome (EM UC7, Leica Microsystems) and mounted on Formvar-coated copper single-slot grid. Sections were stained with 4 % uranyl acetate for 12 m, and lead citrate for 2.5 m at room temperature, then examined using an electron microscope (JEM-1400, JEOL) at 80 kV.

Am, amyloplast; CW, cell wall; ER, endoplasmic reticulum; G, Golgi body; M, mitochondria. Scale bar: 1 μm.

Contributors

Mayuko Sato*, RIKEN Center for Sustainable Resource Science, 1-7-22 Suehiro-cho, Tsurumi-ku, Yokohama, Kanagawa 230-0045, Japan
*E-mail: mayuko.sato@riken.jp

References

10. Staehelin LA, Giddings TH, Kiss JZ, Sack FD (1990) Macromolecular differentiation of Golgi stacks in root tips of *Arabidopsis* and *Nicotiana* seedlings as visualized in high pressure frozen and freeze-substituted samples. Protoplasma 157:75–91

4 The Endoplasmic Reticulum, Golgi Apparatuses, and Endocytic Organelles 79

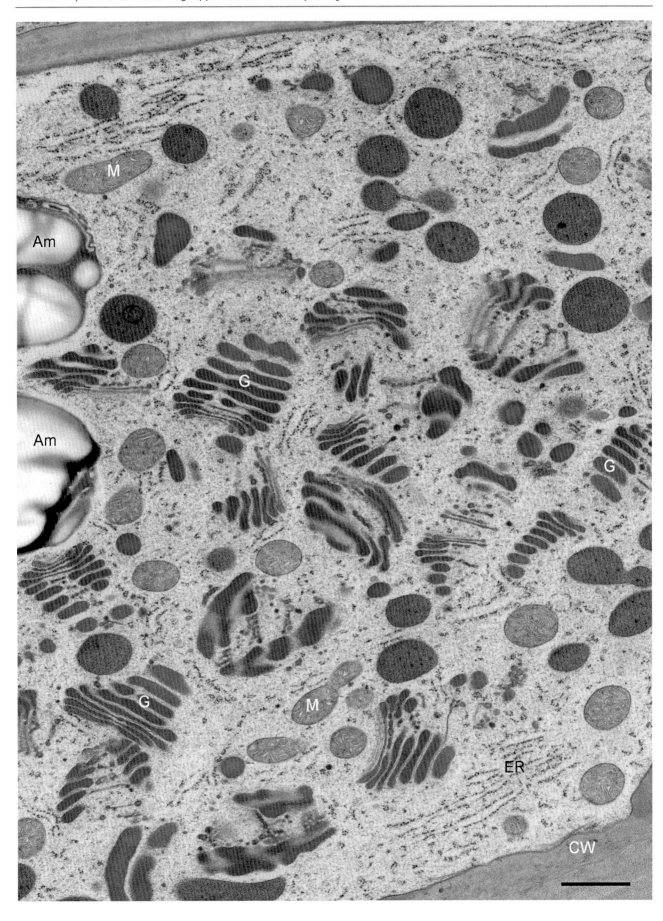

Plate 4.5

Golgi bodies in mature pollen of *Tradescantia reflexa*

In plant and algal cells, Golgi bodies drastically change morphology depending on cell function, but they display similar morphology within a cell [11].

In mature pollen, there are many Golgi bodies composed of four to five cisternae and sometimes a small *trans*-Golgi network with clathrin coated vesicles. These Golgi synthesize cell wall polysaccharides in the cisternae and pack them in Golgi vesicles on the *trans*-side. Golgi vesicles fuse to the plasma membrane and supply membranes and cell wall materials to bud and elongate pollen tubes [12]. The peripheries of *trans*-cisternae are fenestrated, as can be clearly observed in a cisterna cut parallel to its surface (G*). Polysaccharide-synthesizing Golgi bodies that are prominent in active plant cells are not always located near the ER.

This figure shows a transmission electron micrograph of a mature pollen grain of *Tradescantia reflexa* (adapted from [13]). Mature pollen grains in dehiscing anthers were attached to formvar films mounted on wire loops 8 mm in diameter and frozen in liquid propane at -190 °C. Samples were then transferred to acetone (-85 °C) containing 2 % osmium tetroxide and 0.2 % uranyl acetate. After 48 h at -85 °C, samples were gradually warmed to room temperature, washed with acetone, and embedded in Spurr resin. The image was obtained using a transmission electron microscope at 80 kV. ER, endoplasmic reticulum cut perpendicularly to surface; ER*, endoplasmic reticulum cut parallel to surface; G, Golgi body cut perpendicularly to cisternae; G*, Golgi body cut parallel to cisternae. Scale bar: 1 μm.

Contributors

Tetsuko Noguchi*, Course of Biological Sciences, Faculty of Science, Nara Women's University, Kitauoya-nishimachi, Nara 630-8506, Japan
*E-mail: noguchi@cc.nara-wu.ac.jp

References

11. Noguchi T (1978) Transformation of the Golgi apparatus in the cell cycle, especially at the resting and earliest developmental stages of a green alga, *Micrasterias americana*. Protoplasma 95:73–88
12. Noguchi T (1990) Consumption of lipid granules and formation of vacuoles in the pollen tube of *Tradescantia reflexa*. Protoplasma 156:19–28
13. Noguchi T (2009) Golgi apparatus in plant and algal cells. Plant Morphol 21(1):63–70 (Japanese)

4 The Endoplasmic Reticulum, Golgi Apparatuses, and Endocytic Organelles

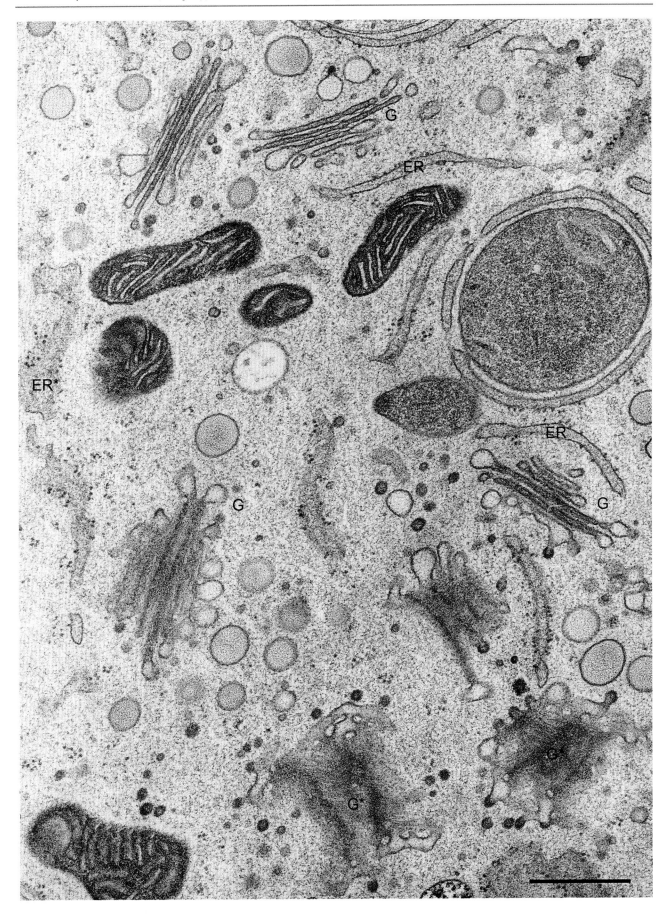

Plate 4.6

Golgi bodies and the *trans*-Golgi networks in *Botryococcus braunii*

To observe the three-dimensional structure of Golgi bodies and *trans*-Golgi networks (TGN), a quick-freeze deep-etch electron microscopy technique is very useful. This technique creates a platinum shadowed carbon replica of the living cell interior.

The green alga *Botryococcus braunii* has large Golgi bodies composed of 7–17 cisternae and a TGN that is the largest known in species ranging from red algae to higher flowering plants. The TGN was first reported in animal cells and defined as a tubular structure with clathrin-coated vesicles which sorts and targets proteins to their appropriate organelles.

Golgi bodies *in B. braunii* always localize near the ER, and their *trans*-side always faces the cell surface. Fenestration of *trans*-Golgi cisternae occurs mainly at their rims. The TGN is composed not only of a tubular structure, but with two domains; the hemispherical TGN-cisterna and TGN-tubules. The TGN cisterna is clearly distinguished from the *trans*-Golgi cisterna by the existence of regularly distributed pores on its surface [14] (The image was adapted from [14]).

Living cells were mounted on copper disks and rapidly frozen in liquid propane ($-190\,^\circ$C). Fracturing, etching and shadowing were carried out at $-110\,^\circ$C and $1\text{--}2 \times 10^{-5}$ Pa. Freeze fracture specimens were etched for 15 min, conically shadowed at a 25° angle with platinum, and subsequently coated with carbon. Replicas were first treated with sodium hypochlorite for 2–3 h, then with 70 % sulfuric acid for 2 days at room temperature to dissolve the underlying cells. Cleaned replicas were mounted on grids and imaged with a transmission electron microscope at 80 kV. CW, cell wall; ER, endoplasmic reticulum; G, Golgi body; TGN, *trans*-Golgi network. Scale bar: 500 nm.

Contributors

Tetsuko Noguchi*, Course of Biological Sciences, Faculty of Science, Nara Women's University, Kitauoya-nishimachi, Nara 630-8506, Japan
*E-mail: noguchi@cc.nara-wu.ac.jp

References

14. Noguchi T, Kakami F (1999) Transformation of *trans*-Golgi network during the cell cycle in a green alga, *Botryococcus braunii*. J Plant Res 112:175–186

4 The Endoplasmic Reticulum, Golgi Apparatuses, and Endocytic Organelles

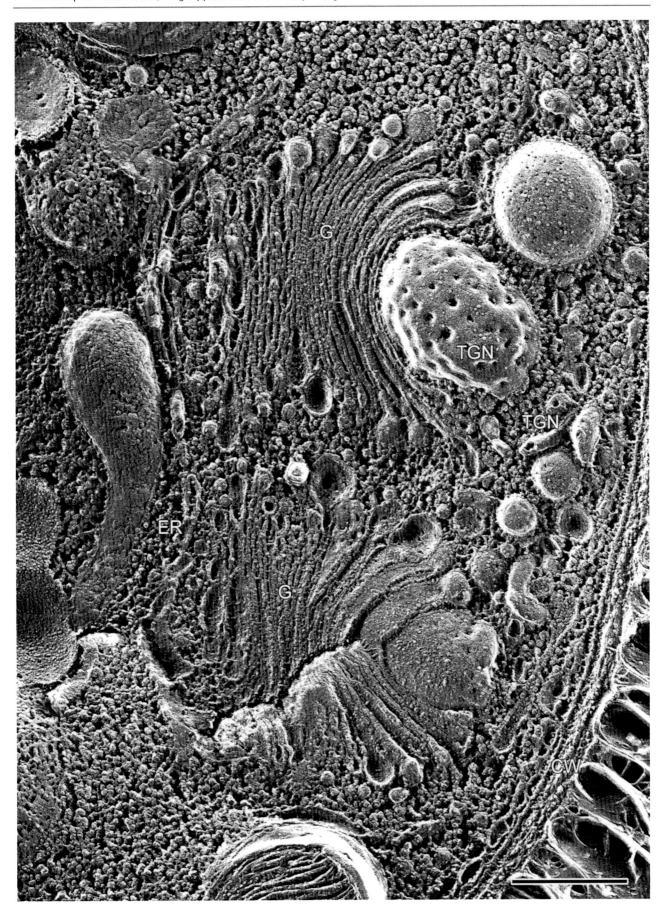

Plate 4.7

Clathrin-coated buds and vesicles in *Botryococcus braunii*

Endocytosis is an energy-consuming process by which cells take in molecules from the extracellular environment and the plasma membrane. Endocytosis occurs in all eukaryotic cells, including plant cells covered with cell walls. In plant cells, the main engulfed substances are large polar molecules that cannot pass through the hydrophobic plasma membrane.

The first step of endocytosis is mediated by two protein complexes, clathrin and an adapter protein complex [15]. The main scaffold component of the coat is the 190 kD clathrin heavy chain and the 25 kD clathrin light chain, which form three-legged trimers called triskelions. The complex plays a structural role by assembling into a basketlike lattice structure on the cytoplasmic surface of the plasma membrane (arrowhead in (**A**) adapted from [16]) that distorts the membrane and drives vesicle budding. The role of the adapter protein complex is to select the specific molecules which will be incorporated into the clathrin bud. The clathrin buds then pinch off from the plasma membrane to form a coated vesicle (arrow in (**A**)). Near the clathrin coated bud, actin filaments (AF), but not microtubules (MT), are prominent.

Multivesicular bodies (MVBs) are unique endosomes containing a number of small vesicles formed by inward budding of the limiting membrane into the lumen (**B**). Plant MVBs appear to play a role in the transport and processing of vacuolar storage proteins and recycling of membrane components removed from the cell plate during cytokinesis.

This figure shows transmission electron micrographs of the cortical regions in *Botryococcus braunii* race A during the growth stage after cell division. Cells attached to formvar films mounted on copper wire loops 8 mm in diameter were frozen in liquid propane at $-190\ ^{\circ}$C. They were then transferred to acetone ($-85\ ^{\circ}$C) containing 2 % osmium tetroxide and 0.2 % uranyl acetate. After 48 h at $-85\ ^{\circ}$C, samples were gradually warmed to room temperature, washed with acetone, and embedded in Spurr resin. AF, actin filament; CW, cell wall; ER, endoplasmic reticulum; G, Golgi body; MT, microtubule, MVB, multi vesicular body; PM, plasma membrane; arrow, clathrin coated vesicle; arrowheads, clathrin coated bud. Scale bars: 500 nm.

Contributors

Tetsuko Noguchi*, Course of Biological Sciences, Faculty of Science, Nara Women's University, Kitauoya-nishimachi, Nara 630-8506, Japan
*E-mail: noguchi@cc.nara-wu.ac.jp

References

15. MacMahon HT, Boucrot E (2011) Molecular mechanism and physiological function of clathrin-mediated endocytosis. Nat Rev Mol Cell Bio 12:517–533
16. Noguchi T (2009) Plant Morphol 21(1)

4 The Endoplasmic Reticulum, Golgi Apparatuses, and Endocytic Organelles

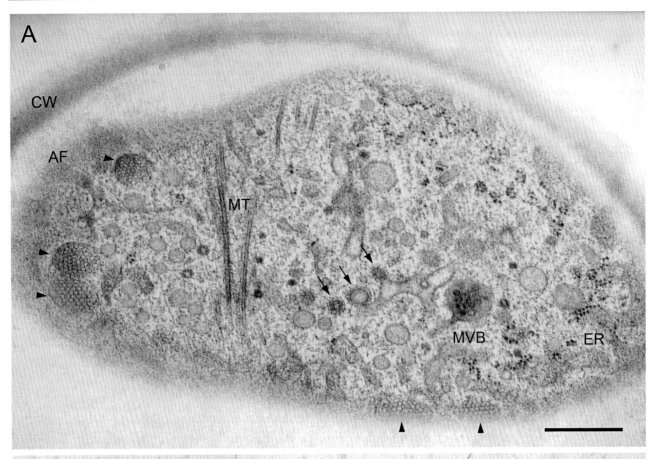

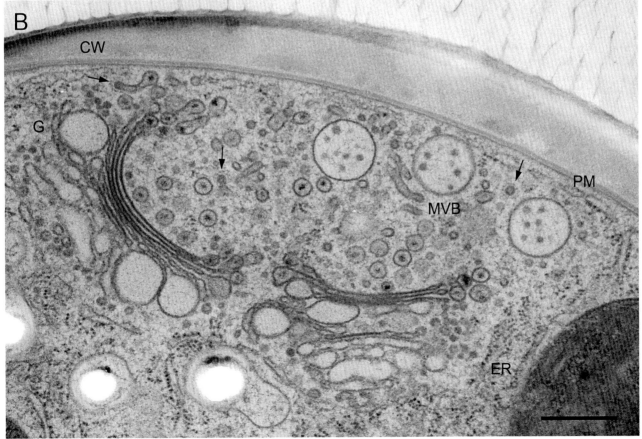

Plate 4.8

Spatio-temporal dynamics of endocytic vesicle formation in *Arabidopsis thaliana*

Endocytosis is a fundamental process for internalizing external materials and regulating the abundance and distribution of proteins and lipids at the plasma membrane (PM). Endocytic vesicle formation involves a series of sequential events, including formation of endocytic vesicles at the PM, recruitment and assembly of coat protein complexes at the PM, membrane invagination followed by membrane scission to form coated vesicles, and removal of coat protein complexes from nascent vesicles. Clathrin is one endocytic coat protein, which is widely conserved in eukaryotic linages. Clathrin coat complexes are composed of clathrin light and heavy chains and adaptor molecules such as cargo- and lipid-binding proteins.

Dynamin-related proteins (DRPs) are multi-domain GTPases that regulate membrane fission, fusion, and tubulation in diverse cellular activities. Among DRP family members, animal dynamin is the best characterized member that participates in clathrin-mediated endocytosis. Dynamin assembles into a helical structure at the neck of the clathrin-coated bud, and constricts in a GTP hydrolysis-dependent manner to sever the bud neck membrane. In contrast to animal cells, two structurally distinct DRPs, DRP1 and DRP2, participate in clathrin-coated vesicle formation at the PM in plant cells [17].

The observation of spatio-temporal behaviors of the endocytosis machinery has helped to elucidate the molecular mechanism of endocytic vesicle formation. Total internal reflection fluorescence microscopy (TIRFM) and related techniques such as variable incidence angle fluorescence microscopy (VIAFM) have been successfully utilized for this purpose, which has allowed high signal-to-noise ratio imaging of fluorescent molecules in a thin surface layer (100–400 nm) adjacent to the cover glass.

The upper panel shows a VIAFM image of a cultured *A. thaliana* cell expressing green fluorescent protein (GFP)-tagged DRP1 (green) and monomeric Kusabira Orange (mKO)-tagged clathrin light chain (red in (**A**)). The lower panel shows the kymograph representing the temporal transition of fluorescence intensities at the bottom edge of the upper figure, which is the first of sequential images taken at 3 s intervals for 3 min (**B**). These images indicate that DRP1 assembles at vesicle formation sites after clathrin assembly, and disappears at the same time as clathrin. The arrow in (**B**) indicates the passage of time.

A cultured *A. thaliana* cell was observed with the Nikon TIRF2 microscopic system (Nikon) equipped with the Cool SNAP HQ2 CCD camera (Roper Scientific, USA). Culture medium containing cells was placed on a slide glass and covered with a coverslip (Matsunami). GFP and mKO were sequentially excited with 488 nm and 561 nm lasers, respectively. Fluorescence emission spectra were separated with a 565LP dichroic mirror and filtered through either a 515/30 or 580LP filter in the Dual View system (Photometrics). Acquired images were processed with Photoshop 7.0 (Adobe Systems), Image Pro Plus 4.0 (Media Cybernetics), and Image J (National Institutes of Health). Scale bar: 2 μm.

Contributors

Masaru Fujimoto[1], Takashi Ueda[2]*, [1]Department of Agricultural and Environmental Biology, Graduate School of Agricultural and Life Sciences, The University of Tokyo, 1-1-1 Yayoi, Bunkyo-ku, Tokyo 113-8657, Japan, [2]Department of Biological Sciences, Graduate School of Science, The University of Tokyo, 7-3-1 Hongo, Bunkyo-ku, Tokyo 113-0033, Japan
*E-mail: tueda@bs.s.u-tokyo.ac.jp

References

17. Fujimoto M, Ueda T (2012) Conserved and plant-unique mechanisms regulating plant post-Golgi traffic. Front Plant Sci 3:197

4 The Endoplasmic Reticulum, Golgi Apparatuses, and Endocytic Organelles

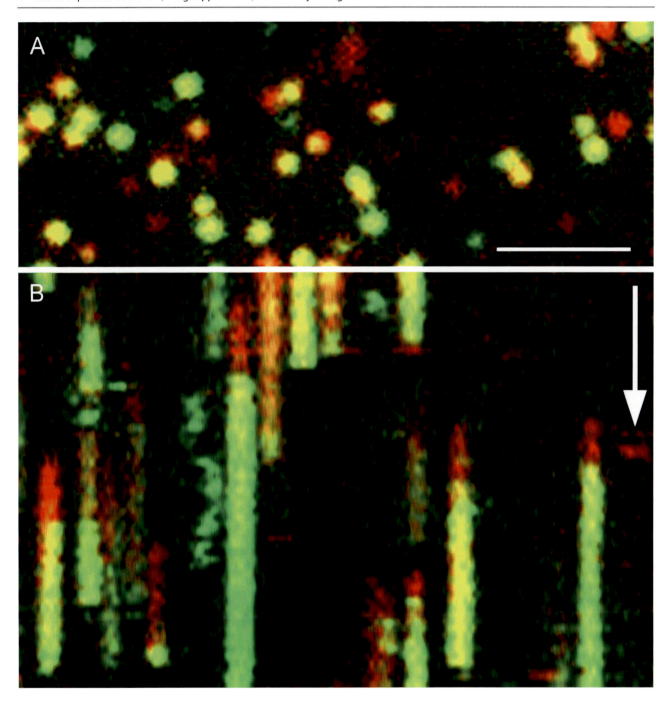

Chapter References

1. Yokota E, Ueda H, Hashimoto K, Orii H, Shimada T, Hara-Nishimura I, Shimmen T (2011) Myosin XI-dependent formation of tubular structures from endoplasmic reticulum isolated from tobacco cultured BY-2 cells. Plant Physiol 156:129–143
2. Ueda H, Yokota E, Kutsuna N, Shimada T, Tamura K, Shimmen T, Hasezawa S, Dolja VV, Hara-Nishimura I (2010) Myosin-dependent endoplasmic reticulum motility and F-actin organization in plant cells. Proc Natl Acad Sci U S A 107:6894–6899
3. Yokota E, Ueda S, Tamura K, Orii H, Uchi S, Sonobe S, Hara-Nishimura I, Shimmen T (2009) An isoform of myosin XI is responsible for the translocation of endoplasmic reticulum in tobacco cultured BY-2 cells. J Exp Bot 60:197–212
4. Farquhar MG, Hauri H-P (1997) Protein sorting and vesicular traffic in the Golgi apparatus. In: Berger EG, Roth J (eds) The Golgi apparatus. Birkhaüser Verlag, Basel/Boston/Berlin, pp 63–129
5. Hirose M, Mukaida F, Okada S, Noguchi T (2013) Active hydrocarbon biosynthesis and accumulation in a green alga, *Botryococcus braunii* (Race A). Eukaryot Cell 12(8):1132–1141
6. Suzuki R, Ito N, Uno Y, Nishii I, Kagiwada S, Okada S, Noguchi T (2013) Transformation of lipid bodies related to hydrocarbon accumulation in a green alga, *Botryococcus braunii* (Race B). PLoS One 8(12):e81626. doi:10.1371/journal.pone.0081626
7. Hayashi Y, Yamada K, Shimada T, Matsushima R, Nishizawa NK, Nishimura M, Hara-Nishimura I (2001) A proteinase-sorting body that prepares for cell death or stresses in the epidermal cells of Arabidopsis. Plant Cell Physiol 42:894–899
8. Matsushima R, Hayashi Y, Kondo M, Shimada T, Nishimura M, Hara-Nishimura I (2002) An endoplasmic reticulum-derived structure that is induced under stress conditions in Arabidopsis. Plant Physiol 130:1807–1814
9. Matsushima R, Kondo M, Nishimura M, Hara-Nishimura I (2003) A novel ER-derived compartment, the ER body, selectively accumulates a beta-glucosidase with an ER-retention signal in Arabidopsis. Plant J 33:493–502
10. Staehelin LA, Giddings TH, Kiss JZ, Sack FD (1990) Macromolecular differentiation of Golgi stacks in root tips of *Arabidopsis* and *Nicotiana* seedlings as visualized in high pressure frozen and freeze-substituted samples. Protoplasma 157:75–91
11. Noguchi T (1978) Transformation of the Golgi apparatus in the cell cycle, especially at the resting and earliest developmental stages of a green alga, *Micrasterias americana*. Protoplasma 95:73–88
12. Noguchi T (1990) Consumption of lipid granules and formation of vacuoles in the pollen tube of *Tradescantia reflexa*. Protoplasma 156:19–28
13. Noguchi T (2009) Golgi apparatus in plant and algal cells. Plant Morphol 21(1):63–70 (Japanese)
14. Noguchi T, Kakami F (1999) Transformation of *trans*-Golgi network during the cell cycle in a green alga, *Botryococcus braunii*. J Plant Res 112:175–186
15. MacMahon HT, Boucrot E (2011) Molecular mechanism and physiological function of clathrin-mediated endocytosis. Nat Rev Mol Cell Bio 12:517–533
16. Noguchi T (2009) Plant Morphol 21(1). (cover)
17. Fujimoto M, Ueda T (2012) Conserved and plant-unique mechanisms regulating plant post-Golgi traffic. Front Plant Sci 3:197

Vacuoles and Storage Organelles

5

Tetsuko Noguchi and Yasuko Hayashi

The vacuole, a membrane-bound water filled organelle which contains inorganic ions and organic compounds, is the most prominent organelle in plant cells. The vacuolar membrane, called the tonoplast, contains transport proteins that maintain cytoplasm homeostasis, including proton pumps (e.g., vacuolar H^+-ATPase and H^+-pyrophosphatase) which stabilize cytoplasmic pH, aquaporins (TIPs) which control water permeability, and Cl^-, Zn^-, Na^- and other ion transporters which selectively transport those ions.

The size and number of vacuoles, as well as the materials they store, varies greatly depending on the cell type and stage of plant development. Mature plant cells contain one central or several very large vacuoles, which restrict the cytoplasm to a small volume between the plasma membrane and the tonoplast. By storing various materials, the central vacuole keeps its water potential as low as that of the cytoplasm, and maintains turgor pressure against the cell wall, which is essential in supporting plants in an upright position.

Plant vacuoles also play roles in molecular degradation and storage to maintain a balance between biogenesis and degradation of many substances and cell structures. In some cells, these functions are carried out by different vacuoles, namely storage vacuoles and lytic vacuoles.

In this chapter, T. Noguchi shows typical central vacuoles in growing *Arabidopsis thaliana* pistil cells using a high-pressure freezing method and transmission electron microscopy. T. Mimura and K. Hamaji demonstrate dynamic changes in vacuole structure in *Arabidopsis* root tip cells under high-salt stress by fluorescence microscopy. Y. Inoue and Y. Moriyasu present autophagosomes in *Arabidopsis* root cells by fluorescence microscopy, and autophagosomes and autolysosomes digesting cytoplasm in BY-2 cells by light and electron microscopy.

In some specialized cells, vacuoles serve as mediators. In protein-seeds, vacuoles store proteins as protein bodies to be used in germination. In oil-seeds and embryos of land plants and some algal cells, lipids are stored in the cytoplasm as oil bodies, which are bounded by a limiting monolayer of phospholipids that is derived from the outer lipid monolayer of membrane of the endoplasmic reticulum. In oil-seeds, lipids stored in oil bodies are transported in peroxisomes (glyoxysomes) and metabolized to produce energy for germination.

Y. Hayashi and S. Mano demonstrate oil metabolism by peroxisomes in *Arabidopsis* cotyledon cells. They show the close relationship between peroxisomes and chloroplasts in a micrograph taken by a laser-scanning confocal microscope, and have captured the direct interaction between oil bodies and glyoxysomes for lipid transport in the etiolated cotyledon by electron microscopy. Y. Kaneko shows the cells of embryonic pea leaves of *Pisum sativum* L containing a large number of protein bodies and lipids using high pressure freezing methods and electron microscopy.

Biofuels have received attention from researchers and industry due to their potential role in combating global warming by minimizing the carbon dioxide emitted by the combustion of fossil fuels. Biofuels include ethanol and methane, which can be generated by fermentation of plant biomass, and neutral lipids such as triacylglycerols and/or hydrocarbons, which can be produced by plants and microalgae. Many photosynthetic algal species have been screened for high lipid content, and some have been characterized as oleaginous and examined for biofuel production capacity. Such organisms could potentially produce 8–24 times more biofuel per unit area than the best land plants.

In this chapter, three algal oil production systems are highlighted. S. Ota and S. Kawano present the dynamic accumulation of oil bodies with astaxanthin during encystment of *Haematococcus pluvialis* by electron microscopy. R. Suzuki and T. Noguchi use fluorescence microscopy to show the green alga *Botryococcus braunii* accumulating large amounts of hydrocarbons in the extracellular space and oil bodies in the cytoplasm, similar to other algae. H. Kuroiwa and T. Kuroiwa have captured the phenomenon of active production of oil bodies in response to nitrogen starvation in the green alga *Chlamydomonas reinhardtii* by fluorescence microscopy and electron microscopy.

T. Noguchi (✉)
Course of Biological Sciences, Faculty of Science, Nara Women's University, Kitauoya-nishimachi, Nara 630-8506, Japan
e-mail: noguchi@cc.nara-wu.ac.jp

Y. Hayashi
Department of Environmental Science and Technology,
Graduate School of Science and Technology, Niigata University,
Ikarashi, Niigata 950-2181, Japan
e-mail: yhayashi@env.sc.niigata-u.ac.jp

T. Noguchi et al. (eds.), *Atlas of Plant Cell Structure*,
DOI 10.1007/978-4-431-54941-3_5, © Springer Japan 2014

Plate 5.1

Central vacuole in *Arabidopsis thaliana* pistil cells

Most mature plant cells contain one central vacuole or several very large vacuoles enclosed by a phospholipid bilayer termed the tonoplast. The central vacuole is filled with water containing inorganic ions and organic compounds, and occupies a large part of the cell's volume, which restricts the cytoplasm to a small volume between the tonoplast and the plasma membrane. In contrast, cells in the meristem lack such a vacuole. As the cell matures, small vacuoles arise from the membrane of the endoplasmic reticulum and Golgi bodies, and they fuse with one another to form a larger vacuole. In addition, fusion occurs among various-sized vacuoles, including between the central vacuole and peripheral vacuoles (instances of all cases are seen in this figure). The central vacuole keeps its water potential as low as that of the cytoplasm, and maintains turgor pressure against the cell wall, which is essential in supporting plant body. It also plays important roles in protoplasmic homeostasis, sequestration of xenobiotics, and digestion of cellular substances and structures. In growing tissues, small vacuolar compartments are distributed around a central vacuole, which are not only pre-vacuoles from the exocytic pathway [1] but also endosomal compartments which arise from the plasma membrane [2].

This figure shows transmission electron micrograph of the upper region of an *Arabidopsis thaliana* L pistil. Pistils were frozen in liquid nitrogen under 2,100 bar pressure in a high pressure freezing apparatus. Frozen cells were transferred to cooled acetone at $-85\,°C$ containing 2 % osmium tetroxide and 0.2 % uranyl acetate. After substitution for 48 h at $-85\,°C$ cells were washed with acetone and embedded in Spurr resin. N, nucleus; V, vacuole. Scale bar: 500 nm.

Contributors

Tetsuko Noguchi*, Course of Biological Sciences, Faculty of Science, Nara Women's University, Kitauoya-nishimachi, Nara 630-8506, Japan
*E-mail: noguchi@cc.nara-wu.ac.jp

References

1. Umekawa M, Klionsky DJ (2012) The cytoplasm-to-vacuole targeting pathway: a historical perspective. Int J Cell Biol. doi:10.1155/2012/142634
2. Neuhaus JM, Martinoia E (2011) Plant vacuoles. Wiley. doi:10.1002/9780470015902.a0001675.pub2

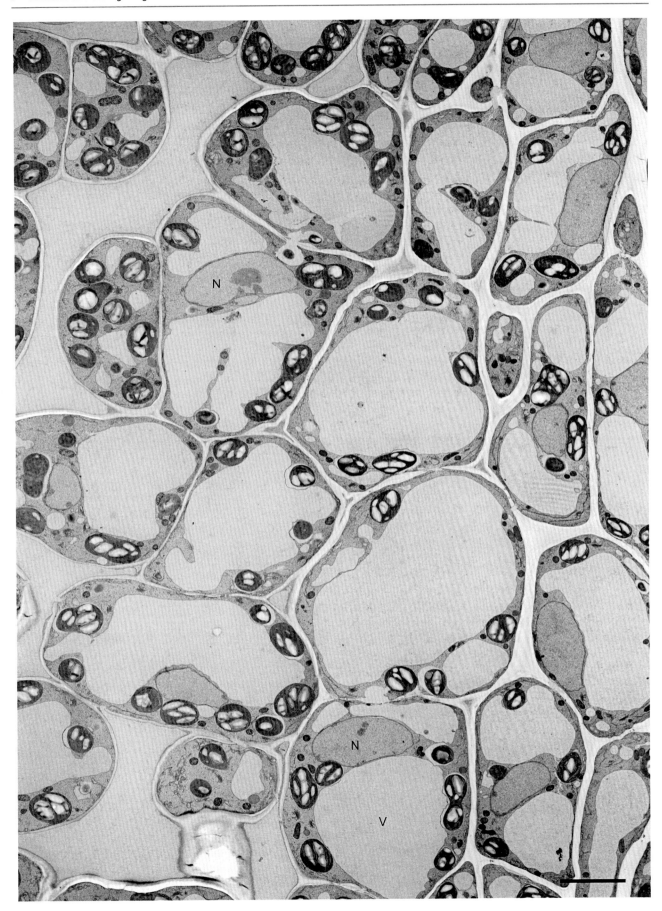

Plate 5.2

Vacuoles under salt stress

The vacuole is the largest and most prominent organelle in a plant cell, and it is enveloped with a single lipid bilayer, the tonoplast. Vacuoles play important roles in various plant cell functions such as space filling, storage of inorganic ions and metabolites, protein degradation, detoxification of heavy metals and other substances, and control of cytoplasmic ionic homeostasis.

In the case of terrestrial plants, high concentrations of Na^+ and Cl^- ions in the cytoplasm are usually toxic. When plants are subjected to high salt conditions, both Na^+ and Cl^- ions are segregated into the vacuole in order to avoid toxicity. The segregation mechanism of the vacuolar membrane involves various vacuolar membrane transporters, and the key transporters for ion homeostasis under salt stress are well-known.

It has been reported that the vacuole changes its structure under salt stress in salt tolerant plants [3, 4]. When these plants were exposed to salt stress, many small vesicles appear and accumulate ions. Subsequently, the vacuolar volume increases. Salt-sensitive plants, such as pea and tomato, do not display rapid increases in vacuolar volume, which suggests that ion sequestration into small vesicles in the initial phase of salt stress may play an important role in tolerance to salt stress.

The figure shows changes in vacuole structure in *Arabidopsis* root tip cells under high salt stress as visualized by fluorescence microscopy [4]. Here, the vacuolar membrane protein VAM3, a SNARE protein functioning in vesicle trafficking, has been transformed with a tagged version, *proVAM3::GFP-VAM3*. Plants (vam3-1) transformed with proVAM3::GFP-VAM3 were established by Dr. Uemura [5]. Control (**A**), and plants were hydroponically treated with 200 mM NaCl for 24 h (**B**). Propidium iodide (PI; 10 µg/mL), a non-penetrating red fluorescent dye, was used to show the cell outlines. Under salt treatment, some cells show strong red fluorescence throughout, suggesting that salt stress diminished membrane integrity. Fluorescent images were obtained with a confocal laser microscope (FV-1000; Olympus) for GFP and PI fluorescence. Scale bars: 30 µm. This figure is adapted from [4].

Contributors

Tetsuro Mimura*, Kohei Hamaji, Department of Biology, Graduate School of Science, Kobe University, Nada, Kobe, Hyogo 657-8501, Japan
*E-mail: mimura@kobe-u.ac.jp

References

3. Mimura T, Kura-Hotta M, Tsujimura T, Ohnishi M, Miura M, Okazaki Y, Mimura M, Maeshima M, Washitani-Nemoto S (2003) Rapid increase of vacuolar volume in response to salt stress. Planta 216:397–402
4. Hamaji K, Nagira M, Yoshida K, Ohnishi M, Oda Y, Uemura T, Goh T, Sato MH, Terao-Morita M, Tasaka M, Hasezawa S, Nakano A, Hara-Nishimura I, Maeshima M, Fukaki H, Mimura T (2009) Dynamic aspects of ion accumulation by vesicle traffic under salt stress in Arabidopsis. Plant Cell Physiol 50(12):2023–2033
5. Uemura T, Morita MT, Ebine K, Okatani Y, Yano D, Saito C, Ueda T, Nakano A (2010) Vacuolar/pre-vacuolar compartment Qa-SNAREs VAM3/SYP22 and PEP12/SYP21 have interchangeable functions in Arabidopsis. Plant J 64(5):864–873

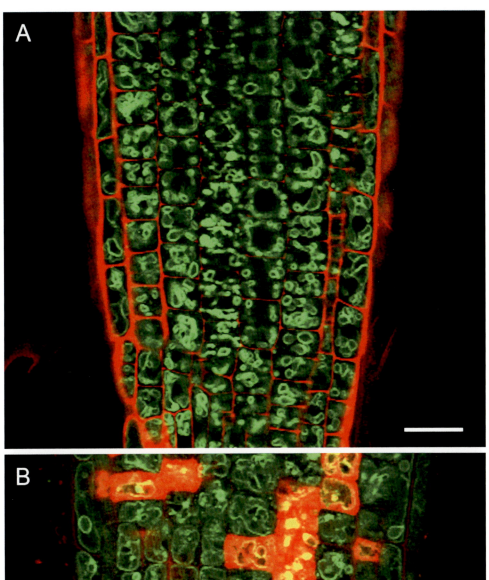
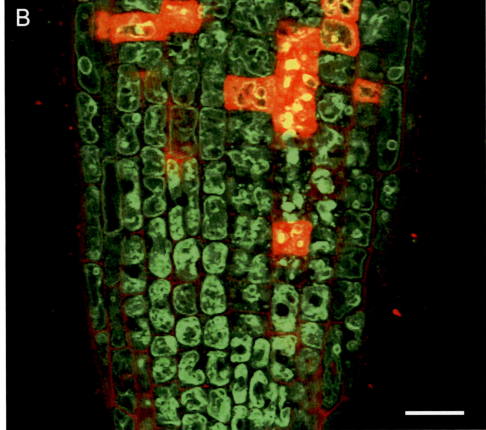

Plate 5.3

Autophagosomes and autolysosomes in plant cells

Autophagy is a phenomenon by which living cells degrade their own constituents in lysosomes and vacuoles. One of the main physiological functions of autophagy is the degradation of cytoplasmic materials for recycling of molecules and supplying substrates for respiration. Autophagy pathways are mainly classified as macroautophagy and microautophagy. In macroautophagy, cytoplasm is first sequestered in autophagosomes, an autophagy-specific organelle. An autophagosome is formed by a double-membrane which engulfs a cellular component. This double-membrane structure is the hallmark of macroautophagy. Autophagosomes subsequently fuse with lysosomes to become autolysosomes in animal cells or with vacuoles in yeast cells. In both cases, the cytoplasm, including any enclosed cellular components, is degraded. The degradation of cytoplasm in vacuoles and lysosomes has been reported in plant cells. In microautophagy, parts of the cytoplasm are directly taken up and degraded by lysosomes or vacuoles.

Autophagosomes can be visualized with tagged Atg8 protein, one of the autophagy-related proteins (Atgs) which localizes to autophagosome double-membranes [6]. In Arabidopsis root cells expressing an Arabidopsis Atg8 (AtAtg8)/green fluorescent protein (GFP) fusion protein, autophagosomes with GFP fluorescence are observed (arrowhead) (**A**). In tobacco BY-2 cells, autophagosomes are induced under sucrose starvation conditions (**B**). Many autophagosomes are observed when BY-2 cells are cultured in sucrose-free medium for 12 h (arrowhead) (**B**). The Nomarski image is on the right (**C**). In BY-2 cells, autolysosomes are also observed (**D, E** adapted from [7]). An autolysosome has a few electron-dense particles, which may be the intermediates of cytoplasmic degradation. Some particles partially retain double-membrane-bound organelle morphology, such as mitochondria or plastids (arrow) (**D, E**).

Arabidopsis seedlings expressing GFP-AtAtg8 fusion protein were grown on 1 % agar plates containing 1/2 Murashige and Skoog (MS) medium and 3 % (w/v) sucrose for 7 days. The root elongation zone was observed with a confocal laser microscope (FV1000-D, Olympus). GFP was excited with a 473 nm laser (LD473, Olympus) (**A**). Tobacco BY-2 cells expressing GFP-AtAtg8 protein were transferred to sucrose-free culture medium and cultured for 12 h. They were observed using a confocal laser microscope (FV1000-D, Olympus) (**B, C**). To induce autolysosome accumulation, BY-2 cells were cultured in sucrose-free culture medium for 14 h in the presence of 10 μM E-64c, a cysteine protease

inhibitor. Cells were fixed with 2 % (w/v) glutaraldehyde and 1 % (w/v) formaldehyde in 0.1 M sodium cacodylate-HCl (pH 6.9) at room temperature for 1 h and at 4 °C overnight and post-fixed with 1 % (w/v) osmium tetroxide at room temperature for 2 h. Samples were then stained en bloc with 1 % (w/v) uranyl acetate for 2 h, the cells were dehydrated and embedded in Spurr resin. Sections were stained with uranyl acetate and lead nitrate, and observed by electron microscopy (H-7000, Hitachi) (**D, E**). Scale bars: 50 μm (**A**), 20 μm (**B, C**), 1 μm (**D, E**).

Contributors

Yuko Inoue, Yuji Moriyasu*, Department of Regulatory Biology, Graduate School of Science and Engineering, Saitama University, Saitama 338-8570, Japan
*E-mail: moriyasu@mail.saitama-u.ac.jp

References

6. Yano K, Suzuki T, Moriyasu Y (2007) Constitutive autophagy in plant root cells. Autophagy 3:360–362
7. Takatsuka C, Inoue Y, Higuchi T, Hillmer S, Robinson DG, Moriyasu Y (2011) Autophagy in tobacco BY-2 cells cultured under sucrose starvation conditions: isolation of the autolysosome and its characterization. Plant Cell Physiol 52:2074–2087

5 Vacuoles and Storage Organelles 95

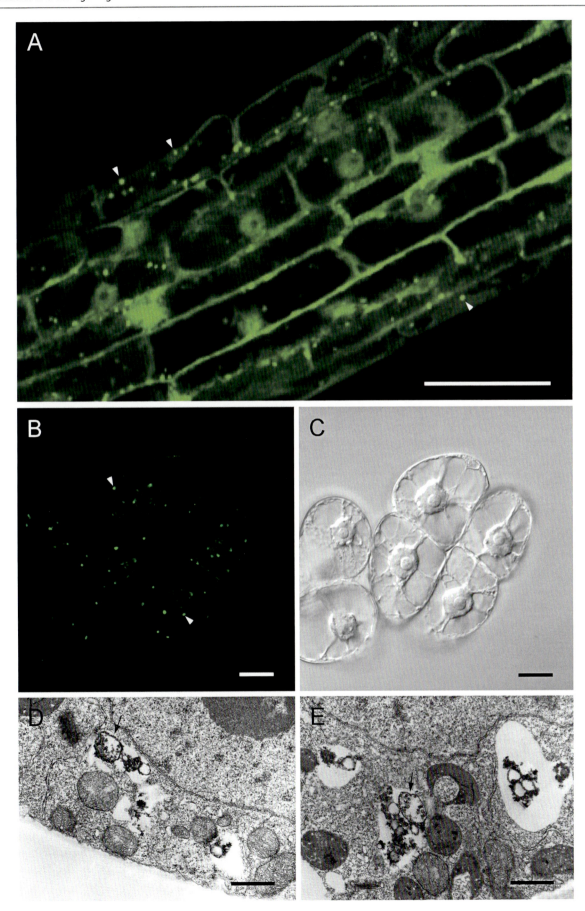

Plate 5.4

Transition of peroxisomes from glyoxysomes to leaf peroxisomes during greening in cotyledon

Peroxisomes can be identified by their single membrane and roughly spherical shape with a diameter of 0.2–1.5 μm. Plant peroxisomes are classified into three groups: glyoxysomes, leaf peroxisomes, and unspecialized peroxisomes. Glyoxysomes are present in oil-seed plant cotyledonary cells, where they play an important role in lipid metabolism during post-germinative growth. Large amounts of triacylglycerols accumulate in organelles called oil bodies during post-germinative growth of the seedlings. The fatty acids released from the triacylglycerols are metabolized via fatty acid β-oxidation and the glyoxylate cycle in glyoxysomes to produce sucrose. Upon exposure to light, etiolated cotyledons become green because they develop chloroplasts and begin to photosynthesize. In these green tissues, leaf peroxisomes are present and play a crucial role in photorespiration, in concert with chloroplasts and mitochondria. In the process of cotyledon greening, glyoxysomes are directly transformed into leaf peroxisomes.

Transgenic *Arabidopsis* cotyledonary cells expressing green fluorescent protein (GFP) with a peroxisomal targeting signal 1 (PTS1) grown in the light for 7 days were examined using a laser-scanning confocal microscope equipped with argon and helium/neon lasers and filter sets for fluorescein (emission filters 505–530 mm and 560–615 mm) [8] (**A**). GFP signals were detected as spherical spots in the transgenic plants. Most peroxisomes are located near chloroplasts. Peroxisomes, chloroplasts and mitochondria are located in close proximity to each other in cotyledons grown in the light, because these organelles act cooperatively in photorespiration, which involves many enzymatic reactions carried out by each organelle (**B**). Glyoxysomes in dark-grown cotyledon are involved in lipid metabolism and as such are surrounded by oil bodies. Wild-type plants have normal glyoxysomes with a uniform electron density under the electron microscope. By contrast, *ped1* mutants with defects in peroxisomal fatty acid β-oxidation show abnormal, large glyoxysomes with tubular structures containing many vesicles (**C**). The tubular structures are made of invaginated peroxisome membranes at the region where they are attached to the oil body. From electron microscopic analysis of serial sections, a mechanism of direct lipid transport from the oil bodies to glyoxysomes during fatty acid β-oxidation was proposed [9]. To observe the structure of this abnormal glyoxysomes clearly, the osmium tetroxide-potassium ferricyanide staining method was used. Scale bars: 10 μm (**A**), 1 μm (**B, C**).

Contributors

Yasuko Hayashi[1]*, Shoji Mano[2], [1]Department of Environmental Science, Graduate School of Science and Technology, Niigata University, Ikarashi, Niigata 950-2181, Japan, [2]Department of Cell Biology, National Institute for Basic Biology, Okazaki 444-8585, Japan
*E-mail: yhayashi@env.sc.niigata-u.ac.jp

References

8. Mano S, Nakamori C, Hayashi M, Kato A, Kondo M, Nishimura M (2002) Distribution and characterization of peroxisomes in Arabidopsis by visualization with GFP: dynamic morphology and actin-dependent movement. Plant Cell Physiol 143:331–341
9. Hayashi Y, Hayashi M, Hayashi H, Hara-Nishimura I, Nishimura M (2001) Direct interaction between glyoxisomes and lipid bodies in cotyledons of the *Arabidopsis thaliana* ped1 mutant. Protoplasma 218:83–94

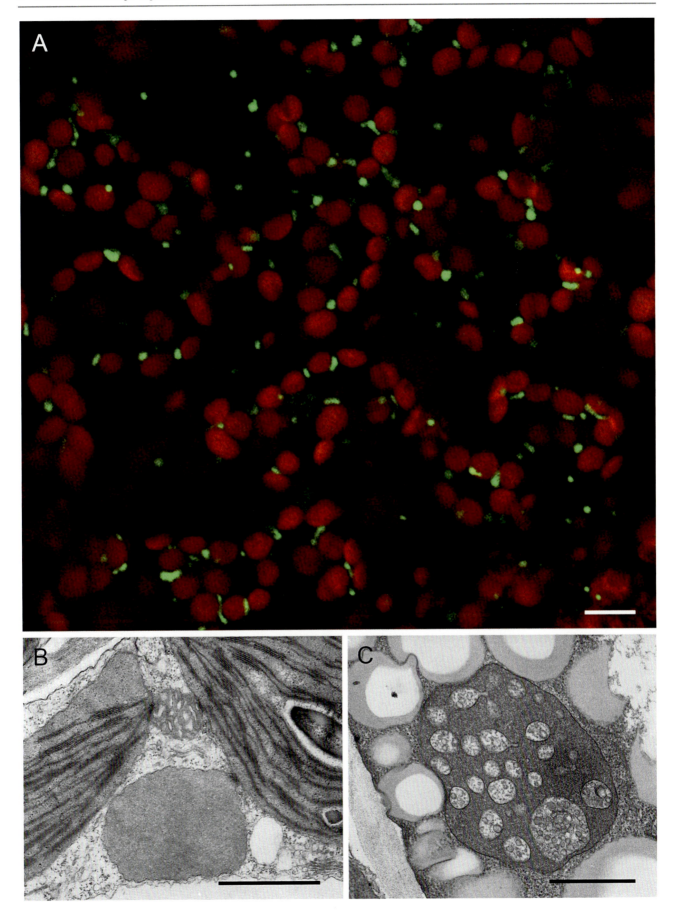

Plate 5.5

Dynamics of embryonic pea leaf cells during early germination

Embryonic pea leaves rapidly develop into functional photosynthetic leaves after imbibition of the dry seeds. This figure illustrates the ultrastructure of third leaves in the plumule of the embryonic axis 6 h after beginning of imbibition. The captured ultrastructure differs greatly from that produced by conventional chemical fixation because samples were prepared by high pressure freezing and freeze substitution [10]. For example, cellular membranes are smoother, organelles appear more turgid, and cytoplasmic and organelle matrices are denser and more homogeneous. Cells have opaque protein bodies and numerous translucent oil bodies lining the plasma membranes. The membranes of endoplasmic reticulum (ER) frequently accompany oil bodies. Many plastids are also covered with ER membranes, and thick bundles of microfilaments are often associated with ER membranes. Most of the oil bodies have retracted from the plasma membranes 16 h after onset of imbibition, and plastid contours become decorated with abundant oil bodies and ER membranes for a short period [11]. Oil bodies surrounding plastids are no longer observed 24 h after onset of imbibition [11]. Oil bodies contents are likely utilized in the development of plastids into functional chloroplasts.

Plumules were excised from pea seeds (*Pisum sativum* L. cv. Perfection) imbibed in water and sliced with a razor blade. Air spaces were eliminated with a drop of 1-hexadecene after the slices (thickness about 0.2–0.4 mm) had been placed in specimen holders. A Balzers HPM 010 high pressure freezer was used to freeze the specimens. Freeze substitution was performed in acetone including 2 % osmium tetroxide at −90 °C overnight, at −80 °C for 5–6 days, at −60 °C overnight and −30 °C overnight. Gradual infiltration with Spurr resin was performed after the specimens had been brought up to room temperature and rinsed in acetone. Ultrathin sections were cut with a diamond knife on a Sorvall MT-2B ultramicrotome and stained with 2 % uranyl acetate (10 min) and then with lead citrate (5 min). A Hitachi H-700H electron microscope was used to observe the sections at an accelerating voltage of 100 kV. Scale bar: 1 μm. This figure is adapted from [12]

References

10. Kaneko Y, Walther P (1995) Comparison of ultrastructure of germinating pea leaves prepared by high-pressure freezing-freeze substitution and conventional chemical fixation. J Electron Microsc 44:104–109
11. Kaneko Y, Keegstra K (1996) Plastid biogenesis in embryonic pea leaf cells during early germination. Protoplasma 195:59–67
12. Kaneko Y (2000) Ultrastructure of germinating pea leaves prepared by high-pressure freezing. Plant Morphol 12(1):10–19 (Japanese)

Contributors

Yasuko Kaneko*, Biology Section in the Faculty of Education, Saitama University, Saitama City, Saitama 338-8570, Japan
*E-mail: yakaneko@mail.saitama-u.ac.jp

5 Vacuoles and Storage Organelles

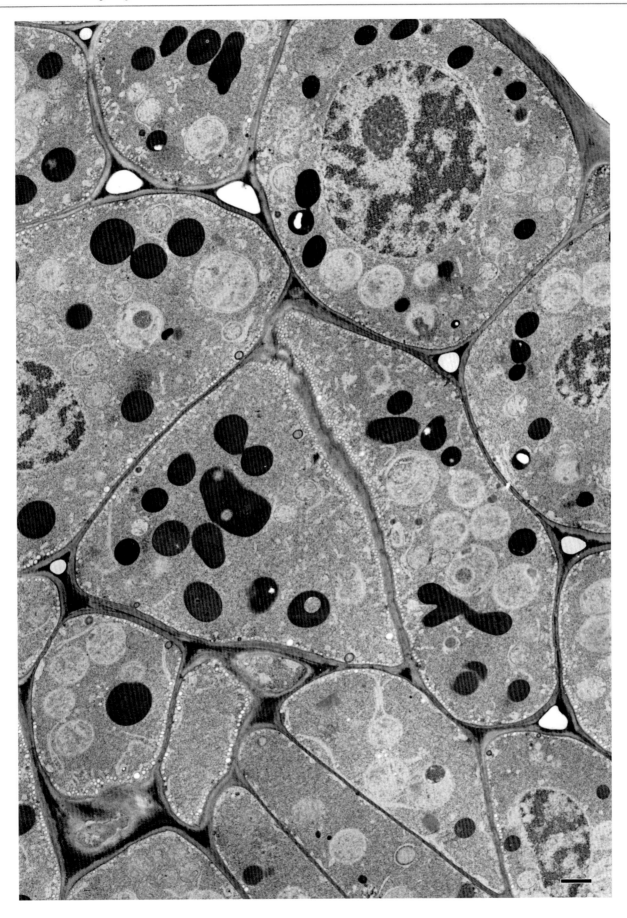

Plate 5.6

Lipids and astaxanthin are major contents of subcellular changes during encystment in *Haematococcus pluvialis*

Haematococcus pluvialis is a freshwater species of green algae which is well known for its production of the strong antioxidant astaxanthin. In the intermediate stage of encystment, living green cells turn greenish-orange due to the accumulation of astaxanthin. In TEM observations, oil bodies (OBs) are moderately electron-dense and locate primarily around the nucleus. OBs are round, of various sizes, and seem to have no membrane-like structure. Fluorescence microscopy indicates that astaxanthin autofluorescence signals colocalize with Nile Red signals, suggesting that astaxanthin coexists with lipids in OBs. This may be the case with cells fixed for TEM imaging, and astaxanthin subcellular distribution may be identified as OBs in transmission electron micrographs. Volumetric analyses [13] show that OBs containing astaxanthin occur throughout cells in the red cyst stage, accounting for approximately half of the total cell volume. Therefore, the most abundant subcellular components during the *Haematococcus* cyst stage may be OBs containing triacylglycerol (TAGs) as the major neutral lipids comprised mostly of palmitic (16:0) and oleic (18:1) acids [14].

This figure is a transmission electron micrograph of a *Haematococcus* cell in the intermediate stage. Cells were fixed with 2.5 % glutaraldehyde and post-fixed with 1 % OsO_4. After the post fixation, cells were dehydrated in a graded ethanol series and incubated with 1:1 ethanol:acetone followed by 100% acetone. Dehydrated samples were infiltrated with increasing concentrations of Supper resin in anhydrous acetone and finally with 100 % Supper resin. Samples were then polymerized at 50 °C for 6 h and 60 °C for 72 h. Ultrathin sections were cut on a Reichert Ultracut S ultramicrotome (Leica) using a diamond knife, and images were obtained using an H-7650 transmission electron microscope at 100 kV. Ch, chloroplast; CW, cell wall; N, nucleus; OB, oil body; P, pyrenoid; SG, starch grain. Scale bar: 5 μm. This figure is adapted from [13].

Contributors

Shuhei Ota*, Shigeyuki Kawano, Department of Integrated Biosciences, Graduate School of Frontier Sciences, The University of Tokyo, Bldg. FSB-601, 5-1-5 Kashiwanoha, Kashiwa, Chiba 277-8562, Japan
*E-mail: ota_shuhei@ib.k.u-tokyo.ac.jp

References

13. Wayama M, Ota S, Matsuura H, Nango N, Hirata A, Kawano S (2013) Three-dimensional ultrastructural study of oil and astaxanthin accumulation during encystment in the green alga *Haematococcus pluvialis*. PLoS One 8:e53618
14. Zhekisheva M, Boussiba S, Khozin-Goldberg I, Zarka A, Cohen Z (2002) Accumulation of oleic acid in *Haematococcus pluvialis* (Chlorophyceae) under nitrogen starvation or high light is correlated with that of astaxanthin esters. J Phycol 38:325–331

5 Vacuoles and Storage Organelles

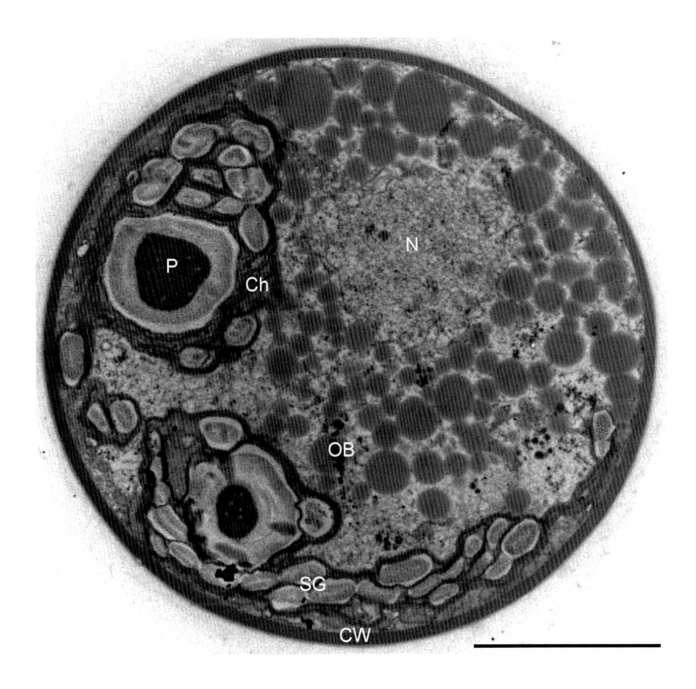

Plate 5.7

Lipid accumulation in the green alga *Botryococcus braunii*

Botryococcus braunii is a colony-forming green alga with pyriform-shaped cells. Among oleaginous microalgae, *B. braunii* is interesting because it accumulates large amounts of hydrocarbons other than triacylglycerols. Moreover, most of the hydrocarbons produced by *B. braunii* accumulate in the extracellular space in addition to the cytoplasm as oil bodies, whereas other microalgae store lipids exclusively in their cytoplasm. *B. braunii* is classified into three principal races (A, B, and L) based on the types of hydrocarbons they synthesize. Race B produces triterpenoids known as botryococcenes and methylsqualenes. Race B has attracted the most attention as an alternative energy source to petroleum because it generally accumulates a higher content of hydrocarbons than the other races, and its hydrocarbons (botryococcenes and methylsqualenes) can be readily converted into biofuels. The botryococcenes and methylsqualenes biosynthetic pathways have been studied, and several important enzymes that contribute to the production of these triterpenoid hydrocarbons have been identified [15]. The increase and decrease of oil bodies in the cytoplasm is strongly related to the accumulation of lipids in the extracellular space just after cell division [16].

This figure shows a fluorescence micrograph of colonies of *Botryococcus braunii* race B stained with the fluorescent lipophilic dye Nile red. Lipids (yellow) accumulate as oil bodies in the cytoplasm and in the intercellular matrix.

One-month old cell cultures were transferred to a fresh culture medium to induce cell division. Cell cultures were first condensed by filtration on 10 μm nylon-mesh on a magnetic filter funnel (PALL), after which the condensed cell culture was centrifuged using Viva-spin ultrafiltration columns at 3,000 rpm for 2 min to remove excess media. Colonies were soaked in 1/100 volume of 1 mM Nile red solution and stained for 10 min, then mounted on a glass slide and strongly covered with a coverslip to align most cells in the focal plane of a fluorescence microscope. Scale bar: 10 μm.

Contributors

Reiko Suzuki, Tetsuko Noguchi*, Course of Biological Sciences, Faculty of Science, Nara Women's University, Kitauoya-nishimachi, Nara 630-8506, Japan
*E-mail: noguchi@cc.nara-wu.ac.jp

References

15. Okada S, Devarenne TP, Murakami M, Abe H, Chappell J (2004) Characterization of botryococcene synthase enzyme activity, a squalene synthase-like activity from the green microalga *Botryococcus braunii*, Race B. Arch Biochem Biophys 422:110–118

16. Suzuki R, Ito N, Uno Y, Nishii I, Kagiwada S, Okada S, Noguchi T (2013) Transformation of lipid bodies related to hydrocarbon accumulation in a green alga, *Botryococcus braunii* (Race B). PLoS One 8(12):e81626. doi: 10.1371/journal.pone.0081626

5   Vacuoles and Storage Organelles   103

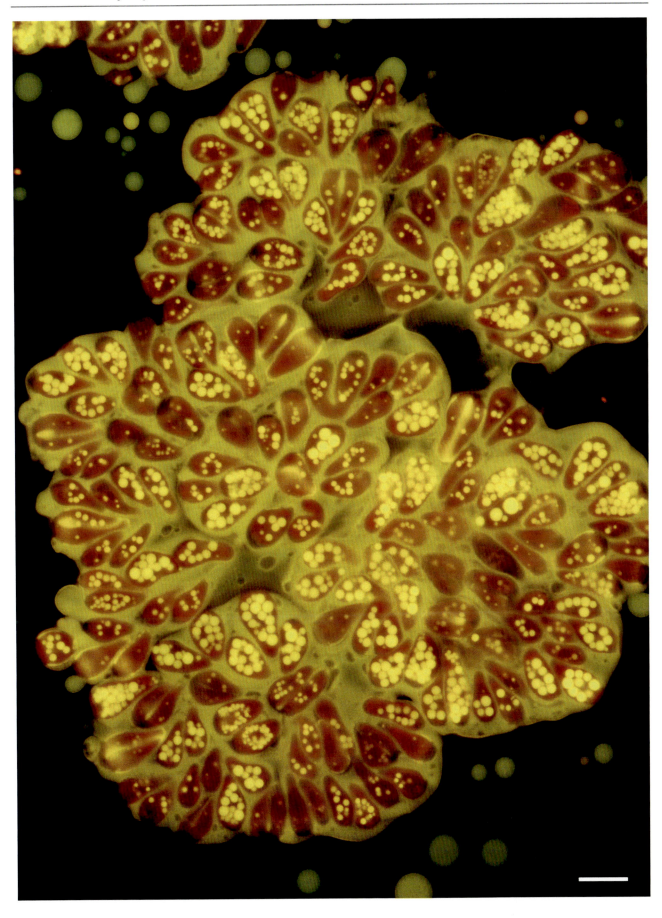

Plate 5.8

Production of oil bodies in response to nitrogen starvation in *Chlamydomonas reinhardtii*

Eukaryotic algae have recently emerged as sources of triacylglycerides (TAGs) which can be converted into diesel and jet transportation fuel. It has reported that under stress conditions many microalgae species accumulate both starch and oil (triacylglycerols) and the model green microalga *Chlamydomonas reinhardtii* produces TAG-filled oil bodies (OBs) in response to nitrogen (N) starvation [17]. Light microscopy and electron microscopy were used to visualize the formation of TAG-filled OBs in *C. reinhardtii* grown in N-starvation media (which, relative to high salt minimum medium, lacks ammonium chloride). Samples were collected every day and stained with BODIPY (4,4-difluoro-3a,4a-diaza-s-indacene), a lipid probe [18]. Spherical yellow ruminous OBs were observed after 3 days (**A**), and with longer culture OBs increased in number and size. After 20 days, the cell was full of enlarged OBs and chlorophyll fluorescence gradually decreased and almost disappeared (**B**).

After 2 days, samples showed several dark masses, which were not OBs, in the lysosome-like organelles. At 3 days, spherical dark OBs appeared in the cytoplasm at the base of the flagella, and after 4 days many OBs were seen in the cytoplasm close to the nucleus and chloroplast, which gradually increased the number. At 20 days, most organelles decreased or disappeared and many large, electron dense OBs merged and occupied nearly the entire cell (**C**). Massive OB production at the expense of own cytoplasm and most organelles can be initiated by N starvation in *C. reinhardtii*.

For electron microscopy, cultured *C. reinhardtii* cells were fixed with 2 % glutaraldehyde and 1 % OsO_4. Scale bars: 1 μm.

Contributors

Haruko Kuroiwa*, Tsuneyoshi Kuroiwa, CREST, Initiative Research Unit, College of Science, Rikkyo University, Toshima, Tokyo 171-8501, Japan
*E-mail: haruko-k@tvr.rikkyo.ne.jp

References

17. Goodson C, Roth R, Wang ZT, Goodenough U (2011) Structural correlates of cytoplasmic and chloroplast lipid body synthesis in *Chlamydomonas reinhardtii* and stimulation of lipid body production with acetate boost. Eukaryot Cell 10(12):1592–1606

18. Kuroiwa T, Ohnuma M, Imoto Y, Misumi O, Fujiwara T, Miyagishima S, Sumiya N, Kuroiwa H (2012) Lipid droplet of bacteria, algae and fungi and relationship between their contents and genome sizes as revealed by BODIPY and DNA staing. Cytologia 77(3):289–299

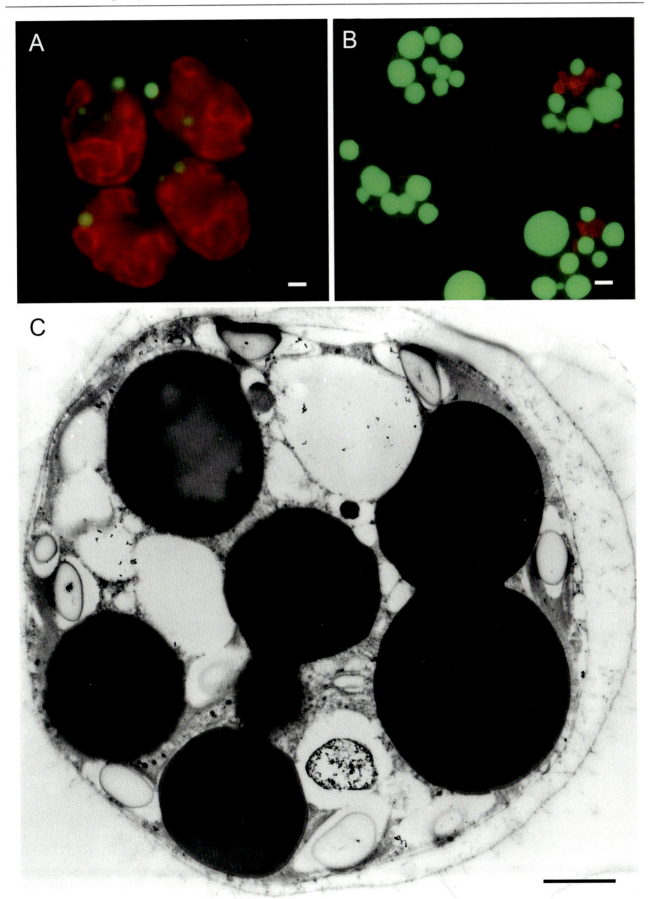

Chapter References

1. Umekawa M, Klionsky DJ (2012) The cytoplasm-to-vacuole targeting pathway: a historical perspective. Int J Cell Biol. doi:10.1155/2012/142634
2. Neuhaus JM, Martinoia E (2011) Plant vacuoles. Wiley. doi:10.1002/9780470015902.a0001675.pub2
3. Mimura T, Kura-Hotta M, Tsujimura T, Ohnishi M, Miura M, Okazaki Y, Mimura M, Maeshima M, Washitani-Nemoto S (2003) Rapid increase of vacuolar volume in response to salt stress. Planta 216:397–402
4. Hamaji K, Nagira M, Yoshida K, Ohnishi M, Oda Y, Uemura T, Goh T, Sato MH, Terao-Morita M, Tasaka M, Hasezawa S, Nakano A, Hara-Nishimura I, Maeshima M, Fukaki H, Mimura T (2009) Dynamic aspects of ion accumulation by vesicle traffic under salt stress in Arabidopsis. Plant Cell Physiol 50(12):2023–2033
5. Uemura T, Morita MT, Ebine K, Okatani Y, Yano D, Saito C, Ueda T, Nakano A (2010) Vacuolar/pre-vacuolar compartment Qa-SNAREs VAM3/SYP22 and PEP12/SYP21 have interchangeable functions in Arabidopsis. Plant J 64(5):864–873
6. Yano K, Suzuki T, Moriyasu Y (2007) Constitutive autophagy in plant root cells. Autophagy 3:360–362
7. Takatsuka C, Inoue Y, Higuchi T, Hillmer S, Robinson DG, Moriyasu Y (2011) Autophagy in tobacco BY-2 cells cultured under sucrose starvation conditions: isolation of the autolysosome and its characterization. Plant Cell Physiol 52:2074–2087
8. Mano S, Nakamori C, Hayashi M, Kato A, Kondo M, Nishimura M (2002) Distribution and characterization of peroxisomes in Arabidopsis by visualization with GFP: dynamic morphology and actin-dependent movement. Plant Cell Physiol 143:331–341
9. Hayashi Y, Hayashi M, Hayashi H, Hara-Nishimura I, Nishimura M (2001) Direct interaction between glyoxisomes and lipid bodies in cotyledons of the *Arabidopsis thaliana* ped1 mutant. Protoplasma 218:83–94
10. Kaneko Y, Walther P (1995) Comparison of ultrastructure of germinating pea leaves prepared by high-pressure freezing-freeze substitution and conventional chemical fixation. J Electron Microsc 44:104–109
11. Kaneko Y, Keegstra K (1996) Plastid biogenesis in embryonic pea leaf cells during early germination. Protoplasma 195:59–67
12. Kaneko Y (2000) Ultrastructure of germinating pea leaves prepared by high-pressure freezing. Plant Morphol 12(1):10–19 (Japanese)
13. Wayama M, Ota S, Matsuura H, Nango N, Hirata A, Kawano S (2013) Three-dimensional ultrastructural study of oil and astaxanthin accumulation during encystment in the green alga *Haematococcus pluvialis*. PLoS One 8:e53618
14. Zhekisheva M, Boussiba S, Khozin-Goldberg I, Zarka A, Cohen Z (2002) Accumulation of oleic acid in *Haematococcus pluvialis* (Chlorophyceae) under nitrogen starvation or high light is correlated with that of astaxanthin esters. J Phycol 38:325–331
15. Okada S, Devarenne TP, Murakami M, Abe H, Chappell J (2004) Characterization of botryococcene synthase enzyme activity, a squalene synthase-like activity from the green microalga *Botryococcus braunii*, Race B. Arch Biochem Biophys 422:110–118
16. Suzuki R, Ito N, Uno Y, Nishii I, Kagiwada S, Okada S, Noguchi T (2013) Transformation of lipid bodies related to hydrocarbon accumulation in a green alga, *Botryococcus braunii* (Race B). PLoS One 8(12):e81626. doi:10.1371/journal.pone.0081626
17. Goodson C, Roth R, Wang ZT, Goodenough U (2011) Structural correlates of cytoplasmic and chloroplast lipid body synthesis in *Chlamydomonas reinhardtii* and stimulation of lipid body production with acetate boost. Eukaryot Cell 10(12):1592–1606
18. Kuroiwa T, Ohnuma M, Imoto Y, Misumi O, Fujiwara T, Miyagishima S, Sumiya N, Kuroiwa H (2012) Lipid droplet of bacteria, algae and fungi and relationship between their contents and genome sizes as revealed by BODIPY and DNA staing. Cytologia 77(3):289–299

Cytoskeletons

6

Ichirou Karahara

The term 'cytoskeleton' might give a static impression given the inclusion of "skeleton". However, the skeletal feature of the cytoskeleton function is only one component of its reality. When a plant cell forms and alters its shape, the cytoskeleton is intimately involved.

The cytoskeleton is comprised of three structural components: microtubules (MTs), actin filaments, and intermediate filaments. The former two play major roles in cellular rearrangements, and their fundamental structures are conserved in animals and plants. The plant cytoskeleton also has its unique functions and organizations. Important roles as well as unique structures of the plant cytoskeleton are outlined in this chapter as depicted at several scales ranging from optical to electron microscopy.

The functions of MTs in mitosis and cytokinesis are the most important because they are crucial to ensure the continuity of life of eukaryotes and contribute to evolution. Y. Mineyuki demonstrates five major MT systems which function during cell cycle progression in *Allium cepa*. Among these systems, interphase cortical MTs, preprophase bands (PPBs), and phragmoplasts are unique to plant cells. Interphase cortical MTs regulate the direction of cell growth. S. Sakaguchi demonstrates the role of MTs in directing the layered structure of shoot apical meristems. Similarly to the mechanism of interphase cortical MT localization in higher plant cells, the unicellular organism *Euglena gracilis* has a unique cytoskeleton, called the pellicle, under the plasma membrane, which is demonstrated by T. Noguchi. Establishment of the direction of cell division is important because cell positions are fixed in plant tissues. The PPB is involved in future division site determination. Actin plays an important role in formation of the PPB. M. Takeuchi and Y. Mineyuki demonstrate an implication of the actin-MT interaction during PPB formation in *Allium cepa*.

The most dramatic function of MTs during cell division is segregating chromosomes to opposite poles in the mitotic apparatus. The organization of mitotic structures of plants is interesting from a phylogenetic viewpoint. We start with a very basic

eukaryotic mitotic apparatus. The MT organizing center (MTOC) in a yeast cell is the spindle pole body, which is functionally equivalent to the centrosome. A. Hirata and S. Kawano present an overview of a spindle pole body in *Saccharomyces cerevisiae*. Algae have centrosomes, including pairs of centrioles, akin to the MTOC in animal cells. C. Nagasato demonstrates spindle formation in a cell of a brown alga, *Scytosiphon lomentaria*. Although flowering plants lack focused MTOCs, liverworts, such as *Pellia endiviifolia*, form polar organizers, which are shaped like centrosomes but differ in that they lack centrioles, as demonstrated by M. Shimamura and Y. Mineyuki. Basal land plants may represent the evolutionary intermediate between algae and flowering plants from the viewpoint of the cell division system. The acentriolar MT system in flowering plants raises a question about how new MTs are generated. T. Murata answers to this question by presenting MT-dependent MT nucleation in tobacco BY-2 cells.

The cytoskeleton is involved in many steps of sexual reproduction. M. Shimamura and Y. Mineyuki discuss MTs at meiosis in bryophytes. C. Nagasato and T. Motomura demonstrate the selective disappearance of female centrioles after fertilization in brown algae, which is similar to the case of paternal centrioles inheritance in animal fertilization. Actin filaments are also involved in sexual reproduction of flowering plants through pollen germination. I. Tanaka demonstrates the distribution of actin filaments in pollen protoplasts of *Lilium longiflorum*. Actin filaments also play important roles related to motility. For example, actin filaments show dynamic changes during spore germination in *Dictyostelium discoideum*. Related to this phenomenon, M. Sameshima demonstrates two unique structures composed of actin which appear in its dormant spores. Despite all that is known, actin filament dynamics are not yet well understood compared to MTs. Live cell imaging with fluorescent probes is a powerful methodology for studying functions of actin filaments. A. Era and T. Ueda visualize the dynamics of actin filaments in the liverwort, *Marchantia polymorpha*. Ultrastructural cytoskeleton imaging is also advancing. Electron tomography, demonstrated by I. Karahara and Y. Mineyuki, is a powerful method for visualizing cellular structures in three dimensions. This method is effective for thick plant samples when employed in conjunction with high-pressure freezing as is demonstrated by I. Karahara et al.

I. Karahara (✉)
Department of Biology, Graduate School of Science and Engineering, University of Toyama, 3190 Gofuku, Toyama 930-8555, Japan
e-mail: karahara@sci.u-toyama.ac.jp

T. Noguchi et al. (eds.), *Atlas of Plant Cell Structure*,
DOI 10.1007/978-4-431-54941-3_6, © Springer Japan 2014

Plate 6.1

Microtubule systems in the cell cycle of onion root tips visualized in 3D

Microtubules (MTs) change their distribution during progression of the cell cycle and participate in morphogenetic changes in plants. Five major MT systems can be seen in somatic cell division in flowering plants. The interphase cortical MT (ICM), which appears in G1, S and part of G2 phases, contributes to cell shape by controlling the direction of cellulose-microfibril deposition. The preprophase band (PPB) in G2 and prophase is involved in the establishment of the division site. Spindle MTs in M phase generate the driving force of chromosome movement. The phragmoplast (PP), which appears in the final stage of cell division, functions in cell plate formation. The radial MT system (RMS) is composed of MTs radiating from the surface of daughter nuclei. Although RMSs are known to be involved in the cell plate positioning in endosperm cytokinesis and in bryophyte meiosis (see Plate 6.5), the RMS is thought to be a transient structure that occurs before the establishment of the ICM in the somatic division of flowering plants [1].

This figure shows a series of stereo-paired MT images in root tip cells of 4 day old onion (*Allium cepa* L. cv. Highgold Nigou) seedlings. Tubulin immunofluoresence images were obtained using a confocal microscope (MRC-500, Japan Bio Rad Co. Ltd, Tokyo, Japan) equipped with a Nikon XF-EFD microscope and a 60x objective (NCF Plan Apo, Nikon, Tokyo, Japan) [2, 3]. MTs run in parallel on the cell cortex and are very few in the cytoplasm in ICM (Adapted from [2] (**A**). Coexistence of a PPB and a spindle in late prophase. MT connections between spindle poles and the PPB are seen. Spindle poles in plants often form 'polar cap' (**B**). Cell in prophase/prometaphse transition stage. Many PPB MTs disappear but a few MTs still remain [3] (**C**). Kinetchore MT fibers show a trunk and numerous branches attached in a fir-tree pattern in metaphase spindles (**D**). Some MT branches cross the midzone and associate with kinetochore MTs in the opposite half spindle. Anaphase spindles with fir-tree (**E**). Kinetochore MTs disappear at the end of anaphase and MTs radiating from the polar region are prominent (**F**). Some MT bundles appear between daughter chromosomes in telophase as cell plate formation begins (**G**). Centrifugally expanding PP with RMS also presents (**H**). Scale bar: 10 μm.

References

1. Mineyuki Y (2007) Plant microtubule studies: past and present. J Plant Res 120:45–51
2. Mineyuki Y, Iida H, Anraku Y (1994) Loss of microtubules in the interphase cells of onion (*Allium cepa* L.) root tips from the cell cortex and their appearance in the cytoplasm after treatment with cycloheximide. Plant Physiol 104:281–284
3. Nogami A, Suzaki T, Shigenaka Y, Nagahama Y, Mineyuki Y (1996) Effects of cycloheximide on preprophase bands and prophase spindles in onion (*Allium cepa* L.) root tip cells. Protoplasma 192:109–121

Contributors

Yoshinobu Mineyuki*, Department of Life Science, Graduate School of Life Science, University of Hyogo, 2167 Shosha, Himeji, Hyogo 671-2280, Japan
*E-mail: mineyuki@sci.u-hyogo.ac.jp

6 Cytoskeletons

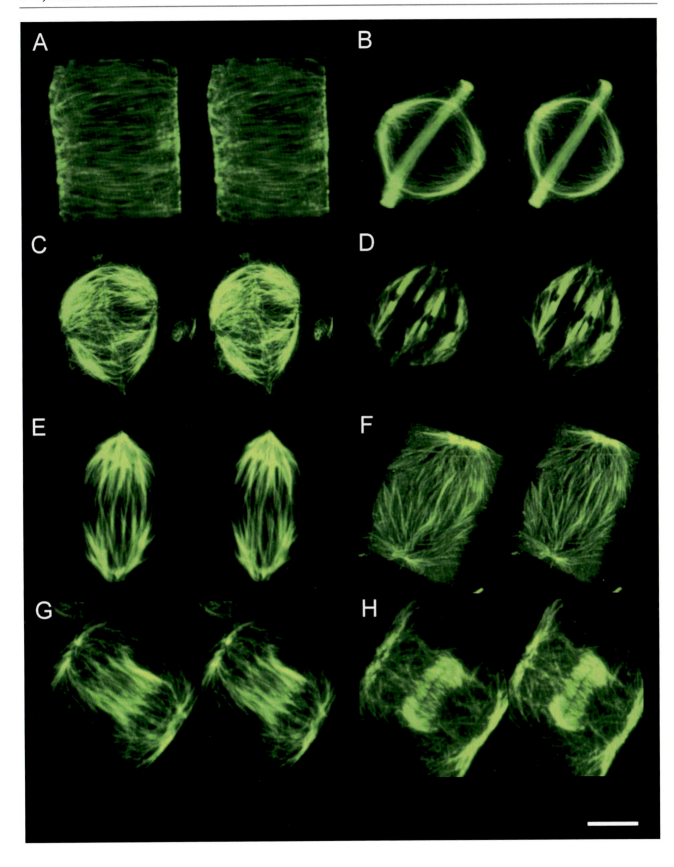

Plate 6.2

Microtubule-dependent microtubule nucleation in a tobacco BY-2 cell

Microtubules (MTs) are a major component of the cytoskeleton. Formation of new MTs, that is, the MT nucleation step, requires polymerization seeds. In living cells, the seeds are γ-tubulin complexes which contain γ-tubulin and other proteins.

In most eukaryotes, the major site of MT nucleation is the centrosome. γ-Tubulin complexes localize to the centrosome and new MTs are formed there. In contrast, vegetative cells of land plants do not have a structurally distinguishable centrosome. Murata et al. have shown that MTs are formed on existing MTs as branches in the interphase cortical MT array [4]. An in vitro MT nucleation reconstruction system also reveals that γ-tubulin plays a role in MT-dependent MT nucleation.

MT-dependent MT nucleation has also been demonstrated to occur in spindle formation of animal cells [5, 6]. The finding of MT-dependent MT nucleation in plant cells will likely contribute several findings to animal cell biology.

MT nucleation in the cortical MT array of a tobacco BY-2 cell is shown (adapted from [4]). The top panel shows GFP-α-tubulin images of MT nucleation (**A**). Cells expressing GFP-α-tubulin were observed by a spinning disk confocal microscope (Olympus IX70 with Yokogawa CSU21) equipped with a 100× oil immersion lens (NA 1.4). Images were acquired with a chilled CCD camera (Coolsnap HQ, Roper). Branch vertices (arrows) and the ends of newly initiated MTs (arrowheads) are shown. The bottom panel shows an electron micrograph of MT branches in an in vitro MT nucleation reconstruction system of MT nucleation (**B**). Protoplasts were attached to a poly-L-lysine coated coverslip, and cytoplasm was removed by shaking in buffer, resulting in plasma membrane-MT complexes being isolated on the coverslip. A cytosolic extract was added onto the coverslip, and the preparation was fixed with glutaraldehyde after 3 min incubation. Sections parallel to the coverslip surface were observed under an electron microscope. Scale bars: 2 μm (**A**), 200 nm (**B**).

Contributors

Takashi Murata*, Division of Evolutionary Biology, National Institute for Basic Biology, Nishigonaka 38, Okazaki, Aichi 444-8585, Japan
*E-mail: tkmurata@nibb.ac.jp

References

4. Murata T, Sonobe S, Baskin TI, Hyodo S, Hasezawa S, Nagata T, Horio T, Hasebe M (2005) Microtubule-dependent microtubule nucleation based on recruitment of γ-tubulin in higher plants. Nat Cell Biol 7:961–968

5. Petry S, Groen AC, Ishihara K, Mitchison TJ, Vale RD (2013) Branching microtubule nucleation in Xenopus egg extracts mediated by augmin and TPX2. Cell 152:768–777

6. Kamasaki T, O'Toole E, Kita S, Osumi M, Usukura J, McIntosh JR, Goshima G (2013) Augmin-dependent microtubule nucleation at microtubule walls in the spindle. J Cell Biol 202:25–33

6 Cytoskeletons

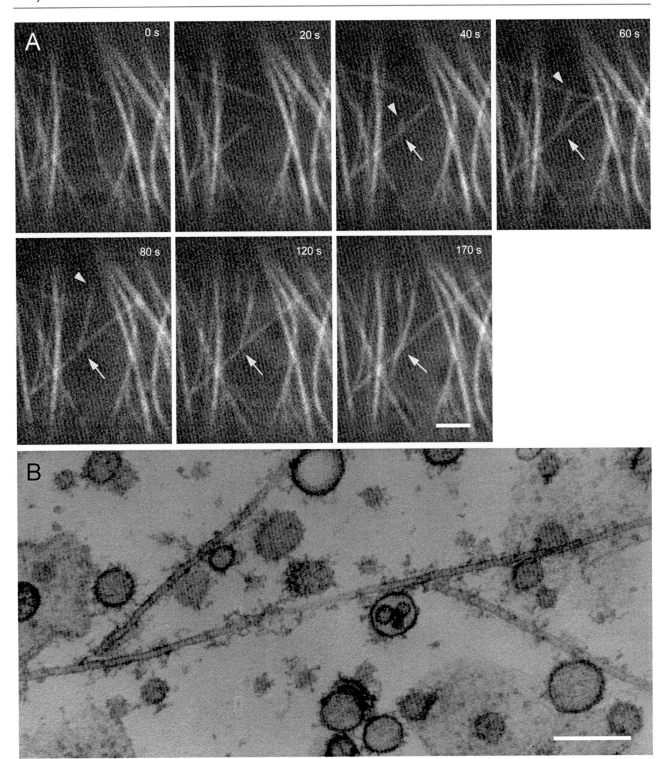

Plate 6.3

Ultrastructural appearance of microtubules in high-pressure frozen onion epidermal cells

Chemical fixatives preserve cells for ultrastructural analysis by immobilizing cellular molecules via crosslinking. This process is slow and only stabilizes subsets of molecules, thereby inducing artifactual changes in cellular architecture. These problems can be overcome by the use of cryofixation methods such as high pressure freezing, which can freeze samples up to 300 μm thick in ~1 ms without introducing structurally damaging ice crystals.

These electron micrographs show images of 40 nm-thick of high-pressure frozen/freeze-substituted onion epidermal cell sections in which microtubules (MTs) are well preserved. The spatial relationship of the cortical MTs (arrows) with the plasma membrane (PM, arrowheads) is clearly seen in a cross-sectional view of an interphase cell (A). The smooth plasma membrane, which exhibits a bilayer architecture, is tightly pressed against the layered cell wall (CW) as a result of turgor pressure, which would not be preserved by chemical fixatives. Cross-sectioned MTs are also visible under higher magnification (A, inset). A dividing epidermal cells in the process of cell plate assembly are visible in longitudinal thin section images (B, C). Cell plate-forming vesicles are associated with phragmoplast MTs (B, arrows), and the MT terminates in the matrix-like structure of the cell plate scaffold within which the vesicles fuse to form the cell plate (adapted from [7]).

Basal cotyledon segments of 3 days old onion (*Allium cepa* L. cv. Highgold Nigou) seedlings were cut with a razor blade while submerged in aqueous solutions of 0.1 M sucrose, and frozen in a BAL-TEC HPM 010 high-pressure freezer (Boeckeler). Frozen specimens were freeze-substituted in 2 % (w/v) OsO_4 in acetone at -80 °C, followed by 2 % (w/v) OsO_4 at 40 °C and 5 % uranyl acetate at 4 °C, then embedded in Spurr resin. Ultrathin sections were stained with uranyl acetate and Reynold's lead citrate, and were imaged with a transmission electron microscope. This freeze substitution protocol has provided both more precise and novel information on the organization of MTs, microfilaments and membrane systems in epidermal cells [8]. Scale bars: 100 nm (A, C), 250 nm (B).

Contributors

Ichirou Karahara[1]*, Takashi Murata[2], Lucas Andrew Staehelin[3], Yoshinobu Mineyuki[4], [1]Department of Biology, Graduate School of Science and Engineering, University of Toyama, 3190 Gofuku, Toyama 930-8555, Japan, [2]Division of Evolutionary Biology, National Institute for Basic Biology, Nishigonaka 38, Okazaki, Aichi 444-8585, Japan, [3]MCD Biology, University of Colorado, Boulder, CO 80309-0347, USA, [4]Department of Life Science, Graduate School of Life Science, University of Hyogo, 2167 Shosha, Himeji, Hyogo 671-2280, Japan
*E-mail: karahara@sci.u-toyama.ac.jp

References

7. Karahara I, Kang BH (2014) High-pressure freezing and low-temperature processing of plant tissue samples for electron microscopy. Methods Mol Biol 1080:147–157
8. Murata T, Karahara I, Kozuka T, Giddings TH Jr, Staehelin LA, Mineyuki Y (2002) Improved method for visualizing coated pits, microfilaments, and microtubules in cryofixed and freeze-substituted plant cells. J Electron Microsc 51:133–136

6 Cytoskeletons 113

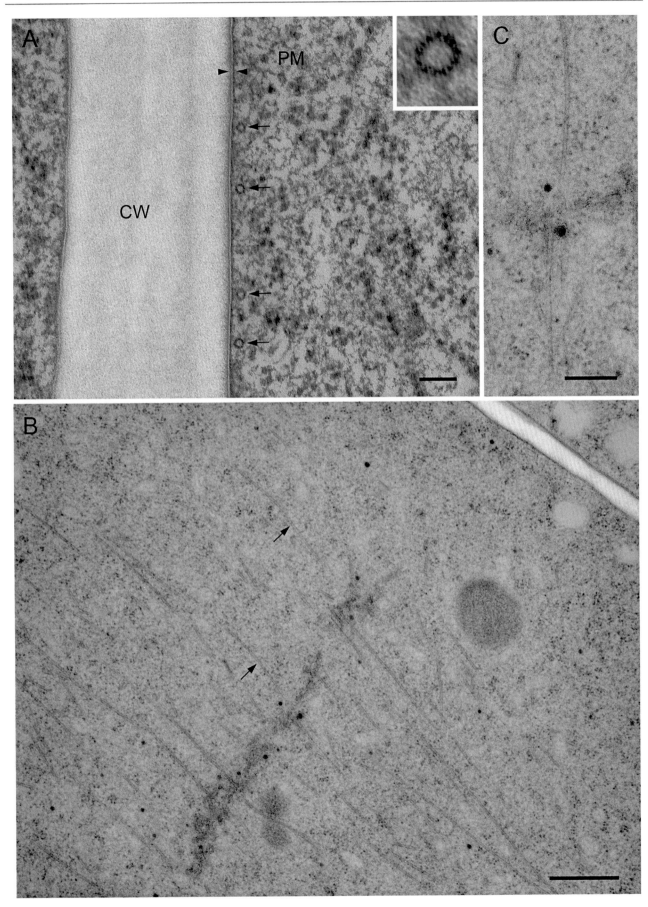

Plate 6.4

Microtubules and their end structures in high-pressure frozen onion epidermal cells visualized by electron tomography

Electron microscopic examination is important to understand how microtubules (MTs) change shape and how they interact with MT-associated structures, although live imaging techniques can reveal the dynamics of single MT. Electron tomography is an emerging technique to visualize and analyze ultrastructural features of cellular and subcellular structures in three dimensions at nanometer-level resolution. This technique is especially effective when employed with cryofixation, which preserves transient endomembrane compartments, MT organization, and coated vesicles. This technique has been used to analyze MT and membrane structure in the preprophase band (PPB), a microtubular nano-machine involved in determination of the future division site of onion cells [9].

This figure shows tomographic slices of a tangentially sectioned onion cotyledon epidermis. The PPB in a late prophase cell is visible in a 1.42-nm-thick tomographic slice (adapted from [9, 10]) (A). The use of electron tomography enables identification, mapping, and modeling of MTs, clathrin-coated pits, and vesicles of PPB regions with a higher degree of resolution than is possible with ultra-thin sections. Four typical of MT end types (arrowheads) are presented here from a collection of tomograms of onion epidermal cells (adapted from [11, 12]) (B). Growing MT ends often have extended formations, and the horned-end is a shrinking MT end. A MT initiates from a γ-tubulin-ring complex and the initiating MT end is capped (yellow arrowhead).

Basal sections of 3 day old onion (*Allium cepa* L. cv. Highgold Nigou) cotyledons were cut, frozen in a high-pressure freezer, freeze-substituted and embedded in Spurr resin. 250 nm-thick outer tangential sections of epidermal cells were mounted in a tilt-rotate sample holder and imaged in a high-voltage microscope operating at 750 kV. Images were taken at x 12,000 from +60° to −60° at 1.5° intervals about two orthogonal axes. A tomogram was computed for each set of aligned tilt series using the R-weighted back-projection algorithm. Tomographic images were presented and analyzed with Imod, the graphics component of the IMOD software package. CCP, clathrin-coated pit; CCV, clathrin-coated vesicle; MT, Microtubule. Scale bar: 1 μm.

Contributors

Ichirou Karahara[1]*, Yoshinobu Mineyuki[2], [1]Department of Biology, Graduate School of Science and Engineering, University of Toyama, 3190 Gofuku, Toyama 930-8555, Japan, [2]Department of Life Science, Graduate School of Life Science, University of Hyogo, 2167 Shosha, Himeji, Hyogo 671-2280, Japan
*E-mail: karahara@sci.u-toyama.ac.jp

References

9. Karahara I, Suda J, Tahara H, Yokota E, Shimmen T, Misaki K, Yonemura S, Staehelin A, Mineyuki Y (2009) The preprophase band is a localized center of clathrin-mediated endocytosis in late prophase cells of the onion cotyledon epidermis. Plant J 57:819–831
10. Karahara I, Suda J, Staehelin LA, Mineyuki Y (2009) Quantitative analysis of vesicles in the preprophase band by electron tomography. Cytologia 74:113–114
11. Mineyuki Y, Suda J, Karahara I (2004) Electron tomography. Plant Morphol 16:21–30 (Japanese)
12. Mineyuki Y (2013) Electron tomography and structure of plant cell framework. In: IIRS (eds) Structure and function of life analyzed by 3D imaging. Asakura Publishing Co. Ltd., Tokyo, pp 51–60. ISBN 978-4-254-17157-0 C3045 (Japanese)

6 Cytoskeletons

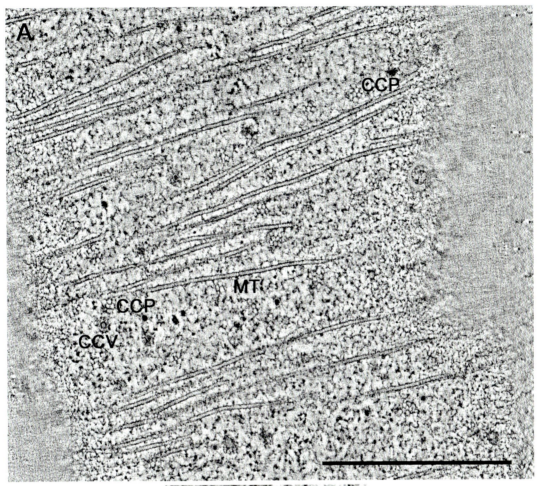

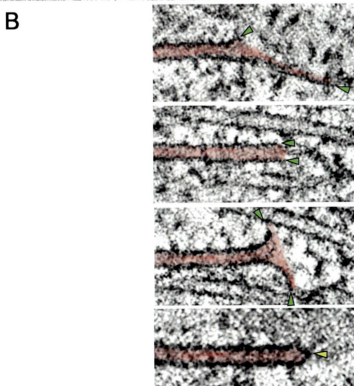

Plate 6.5

Microtubule organizing centers in bryophytes

In animal and algal cells, the typical microtubule organizing center (MTOC) is the centrosome, a compound structure including a pair of centrioles in an amorphous matrix. However, flowering plants have no focused MTOCs comparable to animal centrosomes. Although the evolutionary development of the acentriolar microtubule (MT) system in flowering plants is not well understood, the diversity of the basal land plants MT system may provide clues to the intermediate evolutionary steps from the algal cell division system to the higher plant cell division system [13].

This figure shows triple labeling for tubulin (red signal), γ-tubulin (G9 antibody, green signal) and DNA (DAPI staining, blue signal) in mitotic and meiotic bryophytes cells. In mitotic division in *Pellia endiviifolia* (liverworts), the prophase spindle arises from spherical MTOCs, called polar organizers (POs). POs are shaped like the algal and animal centrosomes but lack centrioles. γ-Tubulin, a ubiquitous component of the MTOC, localizes to POs (**A**). Sporocytes of *Anthoceros punctatus* (hornworts) have a single plastid which divides into four plastids before meiotic nuclear division. MTs arise from the surface of each dividing plastid, which form a quadripolar meiotic spindle. Migration of the four plastid MTOCs establishes the future spore domains and the division polarity of meiosis. In this process, γ-tubulin localizes to the plastid surface (Adapted from [14]) (**B**). In cylindrical *Conocepharum japonicum* (liverwort) sporocytes, incomplete cell plates present at telophase I, but they soon disappear and three cytokinetic division sites are determined in telophase II by a radial MT system (RMS) (see Plate 6.1) emanating from four daughter nuclei. This RMS seems to ensure the equal distribution of cytoplasm between the four spores. During formation of RMS, γ-tubulin locates on the nuclear surface [13] (**C**). Scale bars: 10 μm.

Contributors

Masaki Shimamura[1], Yoshinobu Mineyuki[2]*, [1]Department of Biological Science, Faculty of Science, Hiroshima University, 1-3-1 Kagamiyama, Higashi-Hiroshima, Hiroshima 739-8526, Japan, [2]Department of Life Science, Graduate School of Life Science, University of Hyogo, 2167 Shosha, Himeji, Hyogo 671-2280, Japan
*E-mail: mineyuki@sci.u-hyogo.ac.jp

References

13. Shimamura M, Brown RC, Lemmon BE, Akashi T, Mizuno K, Nishihara N, Tomizawa KI, Yoshimoto K, Deguchi H, Hosoya H, Horio T, Mineyuki Y (2004) γ-Tubulin in basal land plants: characterization, localization, and implication in the evolution of acentriolar microtubule organizing centers. Plant Cell 16:45–59

14. Shimamura M (2004) Monoplastidic cell in lower land plants. Plant Morphol 16:83–92

6 Cytoskeletons 117

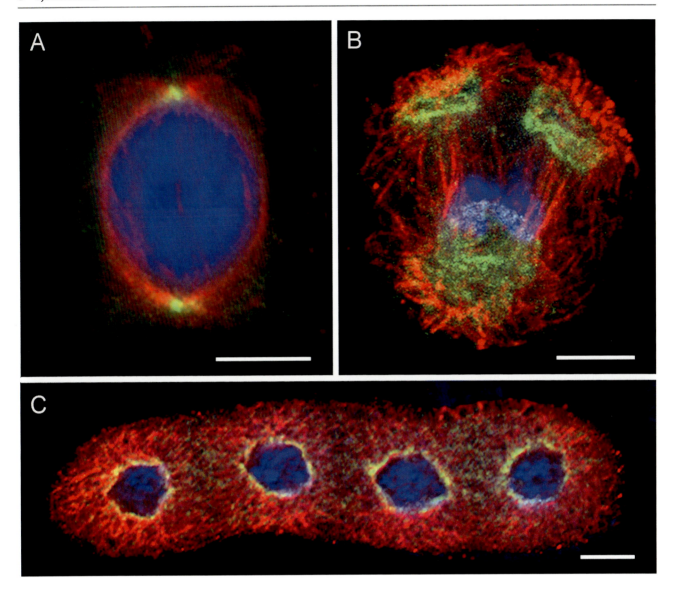

Plate 6.6

Selective disappearance of female centrioles after fertilization in brown algae

Brown algal cells have a definite microtubule (MT) organizing center which is made of a centrosome composed of a pair of centrioles and pericentriolar material. Similar to the case of paternal inheritance of centrioles in animal fertilization (oogamy), it has become clear that centrioles in zygotes of brown algae are paternally inherited, irrespective of the patterns of sexual reproduction, isogamy, anisogamy and oogamy [15]. In oogamy (*Fucus* and *Silvetia*), centrioles in the egg disappear during oogenesis, and the sperm introduces centrioles in the form of flagellar basal bodies.

These figures (adapted from [16]) show the case of anisogamy (*Cutleria cylindrica*). A centrin dot (red) exists at the base of flagella (green) of male and female gametes of *C. cylindrica* (**A, B**). One of the two centrin dots on a zygotic nucleus derived from male and female gametes become gradually disappear (arrow) in 4 h old zygotes (**C**). A disappearing centrin dot is located on a female zygote nucleus in which karyogamy was accidentally delayed (the female nucleus is larger than the male nucleus in anisogamy) (**D**). In isogamy (*S. lomentaria*), one of two centrin dots in male and female zygote gametes disappears. In polyspermic zygotes, one centrin dot always disappears, as is seen in normally fertilized zygotes. Therefore, the maternal centrioles selectively disappear, and paternal centrioles remain, which is to say paternal inheritance of centrioles occurs even in isogamy. TEM observation suggests that degeneration of maternal centrioles begins 1 h after fertilization, the triplet MT cylinder of maternal centrioles shortened from the distal end, and part of the nine MTs triplets became doublets or singlets by degeneration of B and/or C tubules, and in 2 h old zygotes, there is no trace of maternal centrioles.

Gametes and zygotes on coverslips were fixed for 1 h at room temperature in 3 % paraformaldehyde and 0.5 % glutaraldehyde in PHEM buffer. They were treated with 5 % Triton X-100, $NaBH_4$, and blocking solution. After incubation with a polyclonal anti-centrin antibody and a monoclonal anti-β-tubulin antibody, samples were stained with rhodamine-conjugated goat anti-rabbit IgG, FITC-conjugated goat anti-mouse IgG, and DAPI (0.5 µg/ml in PBS). Anti-centrin antibody was used as a centriole marker because centrin is associated with centrioles. Scale bars: 5 µm (**A, B**), 2.5 µm (**C, D**).

Contributors

Chikako Nagasato, Taizo Motomura*, Muroran Marine Station, Field Science Center for Northern Biosphere, Hokkaido University, Muroran, Hokkaido 051-0013, Japan
*E-mail: motomura@fsc.hokudai.ac.jp

References

15. Motomura T, Nagasato C, Kimura K (2010) Cytoplasmic inheritance of organelles in brown algae. J Plant Res 123:185–192
16. Nagasato C, Motomura T, Ichimura T (1998) Selective disappearance of maternal centrioles after fertilization in the anisogamous brown alga *Cutleria cylindrica* (Cutleriales, Phaeophyceae): paternal inheritance of centrioles is universal in the brown alga. Phycol Res 46:191–198

6 Cytoskeletons

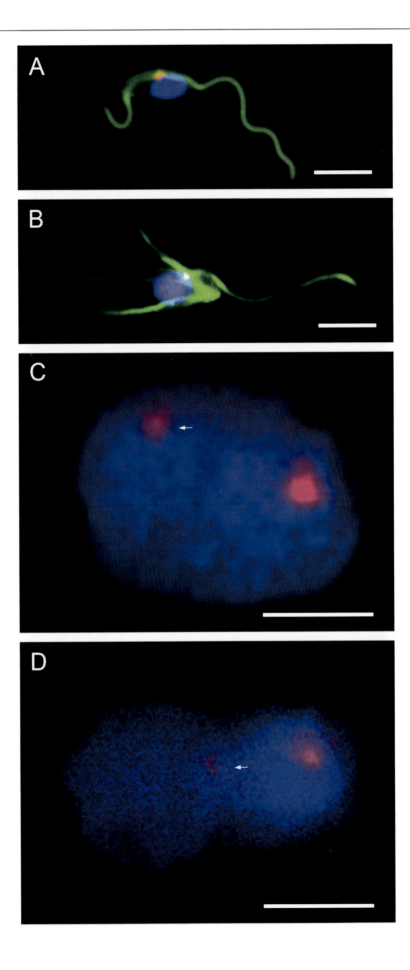

Plate 6.7

Spindle formation in brown algae

Brown algal cells have centrosomes which consist of a pair of centrioles and pericentriolar material. The centrosome exists near the nucleus and acts as the microtubule (MT) organizing center. In brown algal cells, MTs mainly develop from the centrosome. Centriole pairs undergo duplication in S phase, and one pair migrates toward the opposite pole of the nucleus [17]. In prophase, polar MTs extend from each pole, and some pass through a gap of the nuclear envelope [18]. The polar regions of the nuclear envelope develop holes during mitosis, and numerous MTs connect to the chromosomes while most of other regions of the nuclear envelope remain intact throughout mitosis. After mitosis, a pair of centrioles is distributed to each daughter cell.

Images show transmission electron micrographs of metaphase in zygotes of *Scytosiphon lomentaria*. Chromosomes are arranged at the equator of spindle. Centrioles (Ce) can be observed in the both polar regions. Golgi bodies (G) exist in the vicinity of centrioles. The nuclear envelope (arrow) is intact except for the polar region.

For observation by transmission electron microscope, *Scytosiphon* zygotes were rapidly frozen in liquid propane. Frozen samples were transferred to 2 % OsO_4 in anhydrous acetone and stored at $-80°$ C for 2 days. They were gradually warmed to room temperature and washed with acetone. Samples were embedded in Spurr resin. Ultrathin sections were cut and observed under electron microscopy (JEM-1011, JEOL). Scale bars: 1 μm.

Contributors

Chikako Nagasato*, Muroran Marine Station, Field Science Center for Northern Biosphere, Hokkaido University, Muroran, Hokkaido 051-0013, Japan
*E-mail: nagasato@fsc.hokudai.ac.jp

References

17. Motomura T (1994) Electron and immunofluorescence microscopy on the fertilization of *Fucus distichus* (Fucales, Phaeophyceae). Protoplasma 178:97–110
18. Nagasato C, Motomura T (2002) Ultrastructural study on mitosis and cytokinesis in *Scytosiphon lomentaria* zygotes (Scytosiphonales, Phaeophyceae) by freeze-substitution. Protoplasma 219:140–149

6 Cytoskeletons

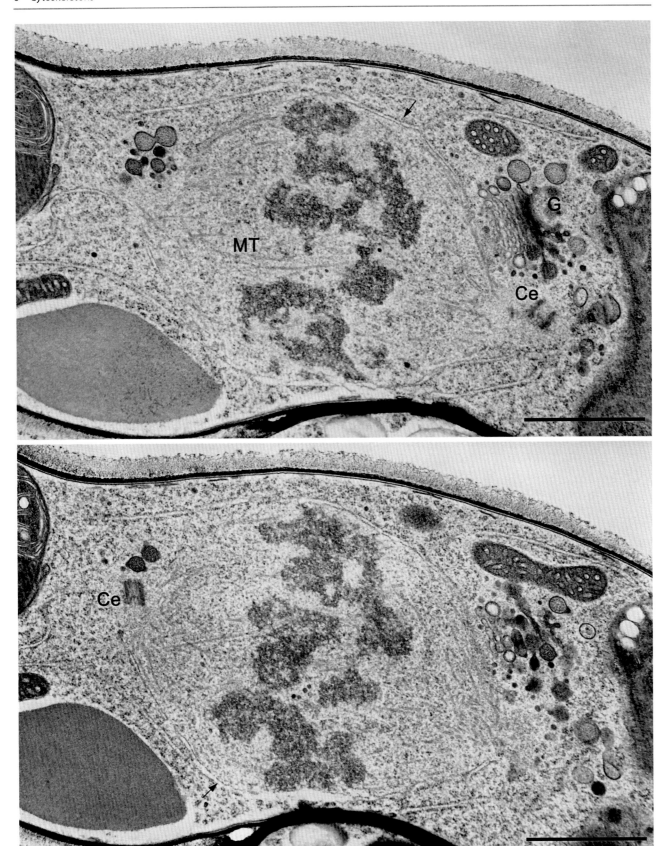

Plate 6.8

Helical rows of microtubules in *Euglena* pellicles

Euglena gracilis is a flagellated unicellular alga characterized as a eukaryotic protist which exhibits animal as well as plant like characters. *E. gracillis* lack cell walls made of cellulose, but have a pellicle on the cell surface. The pellicle is a system of proteinaceous strips oriented along the longitudinal axis of the cell articulated with one another along their lateral margins [19].

E. gracilis cells are deformable exhibit "euglenoid movement," which consists of twisting and wiggling. This movement is instigated by the special structure of pellicle strips which consist of a superficial framework of microtubules (MTs) beneath the plasma membrane. Immunofluorescent microscopy of labelled α-tubulin reveals spiral arrays of pellicles on the cell surface (**inset**) .

Pellicle ultrastructure can be clearly seen by electron microscopy when pellicle strips are transversely sectioned. Under the S-shaped outline of each pellicle notch, 2–5 MT cross sections and the tubules of the characteristic endoplasmic reticulum are always present. Such an arrangement can be seen in the main figure near the central region of the cell surface. In contrast, when pellicle strips are longitudinally sectioned, the flat outline of the cell surface appears. Such an arrangement is apparent in the main figure at the lower region of the cell surface.

For electron microscopy, cells were fixed with 3 % glutaraldehyde and post-fixed with 1 % OsO_4. Post fixation, cells were dehydrated with acetone and embedded in Spurr resin. Ultrathin sections were imaged using a Hitachi H-7000 transmission electron microscope at 80 kV. For immunofluorescent microscopy, cells were fixed with 3.7 % paraformaldehyde and air-dried. Cells were treated by Triton X-100, incubated with a monoclonal anti-α-tubulin and then stained with FITC-conjugated sheep anti-mouse IgG. Scale bars: 500 nm (**main**), 5 μm (**inset**).

Contributors

Tetsuko Noguchi*, Course of Biological Sciences, Faculty of Science, Nara Women's University, Kitauoya-nishimachi, Nara 630-8506, Japan
*E-mail: noguchi@cc.nara-wu.ac.jp

References

19. Yubuki N, Leander BS (2011) Reconciling the bizarre inheritance of microtubules in complex (euglenid) microeukaryotes. Protoplasma 249:859–869

6 Cytoskeletons

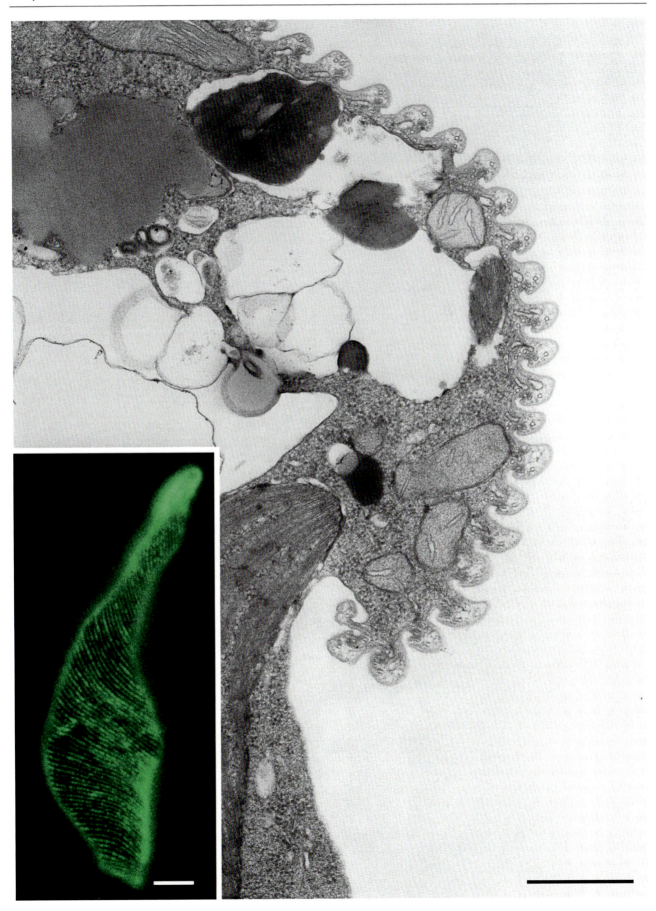

Plate 6.9

Spindle pole body during meiosis I in the budding yeast *Saccharomyces cerevisiae*

The spindle pole body (SPB) is the microtubule (MT) organizing center in yeast cells, functionally equivalent to the centrosome. Unlike the centrosome, the SPB does not contain centrioles. The SPB organizes the MTs which play many roles in the cell. The SPB is important for organizing the spindle, and thus critical to cell division. The SPB in the budding yeast *Saccharomyces cerevisiae* is a complex cylindrical multilayered organelle embedded in the nuclear envelope [20]. The cytoplasmic face of the SPB, or the outer plaque, is next to the intermediate line, central plaque with half-bridge and inner plaque. Meiotic prophase is of particular cytological interest because paired chromosomes are joined along their length by the synaptonemal complex, and the duplicated SPBs, which are joined by the bridge, remain in a side-by-side configuration. SPBs remain paired throughout meiotic prophase, then undergo cleaveage to create half-bridges which separate to form the spindle for meiosis I. This separation requires MTs and results in the generation of a short intranuclear spindle connecting the two SPBs. The spindle is composed of kinetochore MTs and nonkinetochore MTs, which include continuous pole-to-pole and discontinuous MTs. The conclusion of meiosis I is most notable for its failure to produce two separate nuclei, as the original nucleus instead remains single. The two SPBs then undergo another duplication to produce two pairs of duplicated SPBs. Separation of these SPB pairs leads to formation of the two spindles for meiosis II within a single nucleus, which then generates four ascospores.

Cultured *S. cerevisiae* cells were fixed using a rapid freeze fixation and freeze substitution method and observed with TEM. Cultured cells were harvested by centrifugation at 3,300 g for 1 m and rapidly frozen in liquid propane ($-185\,°C$) cooled with liquid nitrogen ($-196\,°C$). Frozen cells were transferred to 2 % OsO_4 in dry acetone at $-80\,°C$ and incubated at $-80\,°C$ for 108 h. Subsequently, samples were gradually warmed from $-80\,°C$ to $0\,°C$ over 5 h, held for 1 h at $0\,°C$, warmed from 0 to $23\,°C$ over 1 h, and incubated at $23\,°C$ for 2 h. Samples were washed three times with dry acetone at room temperature and then infiltrated with increasing concentrations of Spurr resin in dry acetone and finally with 100 % Spurr resin. Ultrathin sections (0.06–0.07 μm) were cut with a diamond knife on an ultramicrotome and mounted on Formvar-coated copper grids. Sections were stained with 3 % uranyl acetate for 2 h at room temperature, with lead citrate for 10 m at room temperature and examined with an electron microscope at 100 kV. This figure is adapted from [21]. Scale bar: 250 nm.

Contributors

Aiko Hirata[1], Shigeyuki Kawano[2]*, [1]Bioimaging Center and [2]Department of Integrated Biosciences, Graduate School of Frontier Sciences, The University of Tokyo, Bldg. FSB-601, 5-1-5 Kashiwanoha, Kashiwa, Chiba 277-8562, Japan
*E-mail: kawano@k.u-tokyo.ac.jp

References

20. Jaspersen SL, Winey M (2004) The budding yeast spindle pole body: structure, duplication, and function. Annu Rev Cell Dev Biol 20:1–28
21. Hirata A (2010) Technical note: Meiosis I in *Saccharomyces cerevisiae* by rapid-freeze electron microscopy. Cytologia 75:221–222

6 Cytoskeletons

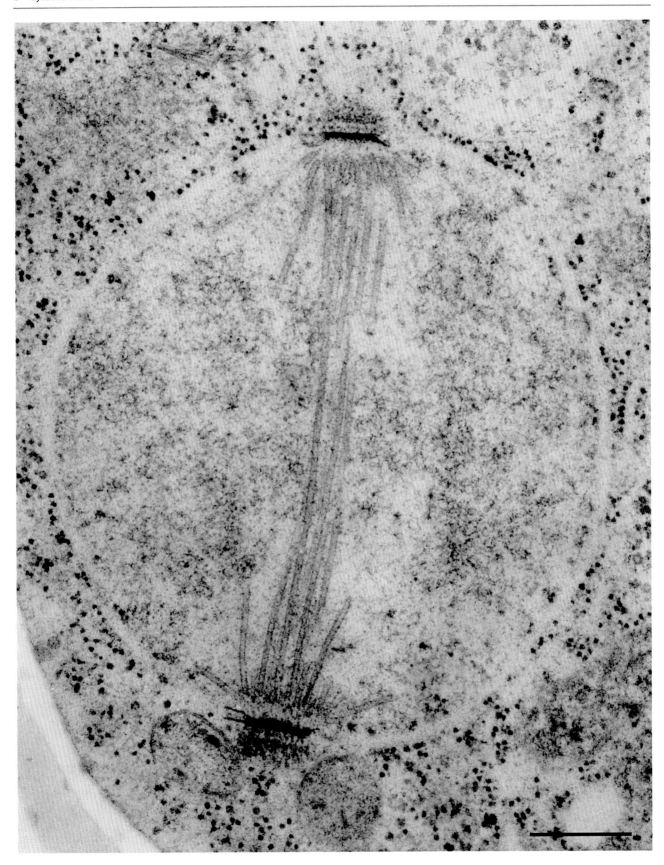

Plate 6.10

Actin filaments in *Lilium longiflorum* pollen protoplasts

In flowering plant sexual reproduction, pollen plays a crucial role in delivering male gametes to the embryo sac through the elongated pollen tube. Actin filaments play an important role in the process of pollen germination and pollen-tube growth because cytochalasins promptly inhibit tube growth and cytoplasmic streaming. Although actin dynamics can be visualized during germination with rhodamine-phalloidin staining, the visualization of actin filaments in ungerminated pollen is generally difficult because of the presence of a thick pollen wall which includes exine. To overcome this difficulty, pollen protoplasts can be used. Intact protoplasts can be directly released from maturing lily pollen through a break in the aperture devoid of sculptured exine [22]. The composition of the enzyme mixture is simple, 1 % macerozyme and 1 % cellulase, and pollen protoplasts can be cultured through normal pollen-tube development for a prolonged period.

This figure shows a freshly isolated pollen protoplast of *Lilium longiflorum*, directly stained with 0.33 μM rhodamine-phalloidin and observed under epifluorescence microscope (adapted from [23]) (**A**). Minute actin filaments which form a fine network throughout the cytoplasm of the vegetative cell are clearly visible. Because the profile resembles that of activated *Lilium* pollen, it is assumed that enzyme treatment may induce pollen activation by simultaneous hydration. In culture, cortical actin filaments gradually organize and align to form thick bundles. Germination occurs parallel to the ordered array of actin bundles (**B**). Numerous actin filaments are seen within the germinated pollen tube. There is an intimate spatial relationship between the arrangement of actin and pollen germination. Further, the organization and arrangement of actin is sensitive to cytochalasin B, and germination is completely blocked. GC, Generative cell; VC, vegetative cell; N^{-GC}, nucleus of generative cell; N, nucleus of vegetative cell; PT, pollen tube. Scale bars: 20 μm.

Contributors

Ichiro Tanaka*, Graduate School of Nanobioscience, Yokohama City University, Seto 22-2, Kanazawa-ku, Yokohama, Kanagawa 236-0027, Japan
*E-mail: itanaka@yokohama-cu.ac.jp

References

22. Tanaka I, Kitazume C, Ito M (1987) The isolation and culture of lily pollen protoplasts. Plant Sci 50:205–211
23. Tanaka I, Wakabayashi T (1992) Organization of the actin and microtubule cytoskeleton preceding pollen germination: an analysis using cultured pollen protoplasts of *Lilium longiflorum*. Planta 186:473–482

6 Cytoskeletons 127

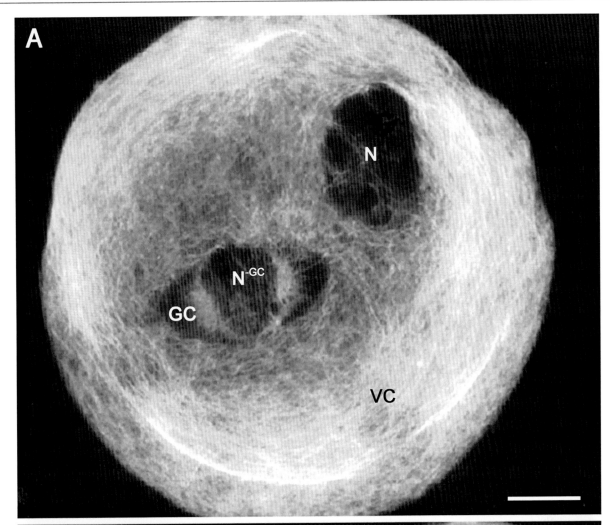

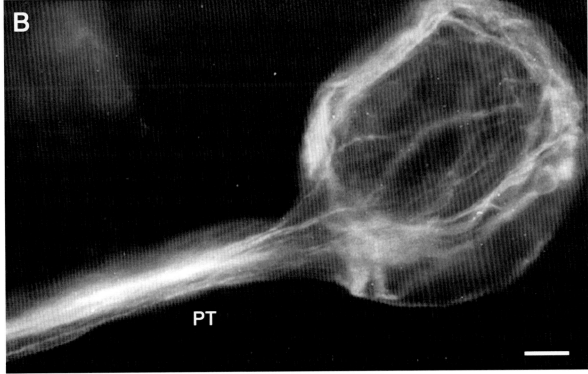

Plate 6.11

Dynamics of actin filaments in the liverwort, *Marchantia polymorpha*

In plants, actin filaments play crucial roles in multiple cellular processes, including cytoplasmic streaming, organelle movement, cell morphogenesis, cell division, and tip growth. Live cell imaging with fluorescent probes provides a powerful methodology for studying these important functions of actin filaments. Several probes have been developed for visualizing actin filaments. Lifeact, a peptide comprising the first 17 amino acids of the yeast protein Abp140p, has been shown to function as such a probe [24]. Lifeact peptide tagged with Venus (modified YFP) was used to visualize actin filaments in plant cells including a model bryophyte, *Marchantia polymorpha*. In transgenic plants expressing Lifeact-Venus, visualized actin filaments further reveal that actin filaments actively move in *M. polymorpha* cells [25]. Directional sliding of actin filaments is a unique feature of actin behavior in *M. polymorpha*. This unique sliding movement was also demonstrated to be accelerated by stabilization or depolymerization of microtubules (MTs), suggesting that MTs have a regulatory role in actin filament motility in *M. polymorpha* [26]. Actin filaments are frequently and consistently observed to be associated with MTs in *M. polymorpha* cells.

This picture shows sliding actin filaments labeled by Lifeact-Venus in a MT-stabilized *M. polymorpha* cell (adapted from [26]). Young thalli of transgenic *M. polymorpha* expressing Lifeact-Venus under the control of the Cauliflower mosaic virus 35S promoter were treated with 10 μM paclitaxel for 2 h at room temperature to stabilize MTs. Samples were then observed under a fluorescent microscope (model BX51; Olympus) equipped with a confocal scanner unit (model CSU10; Yokogawa Electric) and a cooled CCD camera (model ORCA-ER; Hamamatsu photonics). Sequential images taken every 3 s were colored red, green, and blue, and superimposed using ImageJ software (http://rsb.info.nih.gov/ij/). Sliding actin filaments are shown as colorful lines with gradation along the filament long axis from red to blue. Scale bar: 10 μm.

Contributors

Atsuko Era, Takashi Ueda*, Department of Biological Sciences, Graduate School of Science, The University of Tokyo, 7-3-1, Hongo, Bunkyo-ku, Tokyo 113-0033, Japan
*E-mail: tueda@bs.s.u-tokyo.ac.jp

References

24. Riedl J, Crevenna AH, Kessenbrock K, Yu JH, Neukirchen D, Bista M, Bradke F, Jenne D, Holak TA, Werb Z, Sixt M, Wedlich-Sordner R (2008) Lifeact: a versatile marker to visualize F-actin. Nat Methods 5:605–607
25. Era A, Tominaga M, Ebine K, Awai C, Saito C, Ishizaki K, Yamato KT, Kohchi T, Nakano A, Ueda T (2009) Application of lifeact reveals F-actin dynamics in *Arabidopsis thaliana* and the liverwort, *Marchantia polymorpha*. Plant Cell Physiol 50:1041–1048
26. Era A, Kutsuna N, Higaki T, Hasezawa S, Nakano A, Ueda T (2013) Microtubule stability affects the unique motility of F-actin in *Marchantia polymorpha*. J Plant Res 126:113–119

6 Cytoskeletons

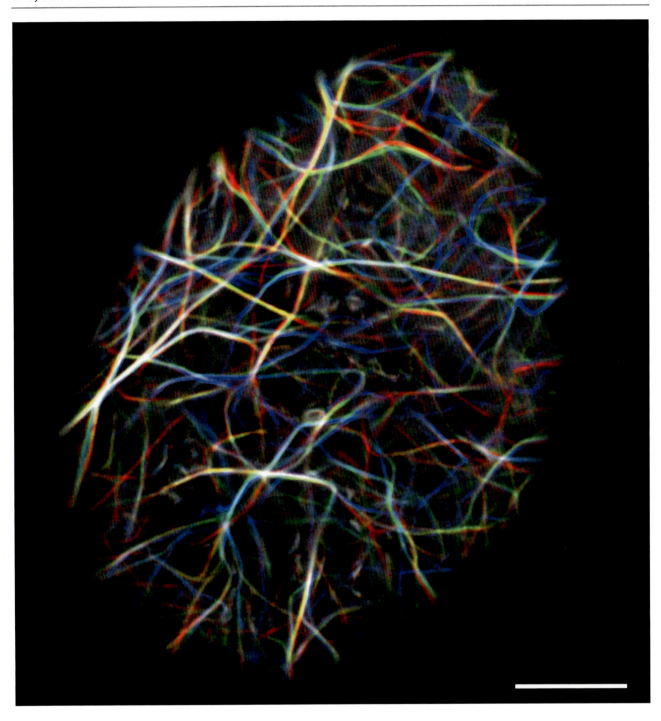

Plate 6.12

Two actin structures in dormant *Dictyostelium discoideum* spores

Upon food source exhaustion, vegetative *D. discoideum* cells aggregate to become multicellular and differentiate into dormant spores and stalk cells in a fruiting body. Morphologically dormant spores are static, indicating that the actin cytoskeleton is generally inactive. In spores, half of the actin molecules are tyrosine-phosphorylated, and maintenance of high levels of actin phosphorylation correlates with the inactive state of the actin cytoskeleton. In addition to actin phosphorylation, two types of structures composed of either actin filaments (adapted from [27]) (**A–C**) or G-actin (adapted from [28]) (**D**) appear in the spores.

In actin filaments, three actin filaments are bundled to form a tubular structure (**C**) which are then organized into rods [27]. These tubular structures are 13 nm diameter are termed actin tubules. The rods first appear in premature spores at the mid culmination stage as modules of short bundles of actin tubules are hexagonally cross-linked (**B**). Formation of actin rods begins late in the culmination stage and proceeds until two days after completion of fruiting bodies [29]. Physical events include association of several modules of bundles, close packing and a decrease in the diameter of actin tubules, and elongation of rods across the nucleus or the cytoplasm. Actin phosphorylation levels increase and reach maximum level before completion of actin rod elongation. Immediately following activation of spore germination, actin is rapidly dephosphorylated, followed shortly thereafter by the disappearance of rods. Rod maturation processes may correlate with actin phosphorylation. S-adenosyl-L-homocysteine hydrolase (SAHH) only accumulates with actin rods at the spore stage of the life cycle. SAHH is believed to be a target for antiviral chemotherapy and the suppression of T cells. Cofilin is also a component of cytoplasmic rods, but few cofilin molecules are included in nuclear rods.

Another structure composed of actin is an aggregate of G-actin (**D**) surrounded by lipid droplets [28]. Anti-actin antibody signals are observed substantially in the electron dense area when thin sections of freeze-substituted spores are immuno-stained, indicating that actin molecules accumulate densely in this area. However, nonfibrous structure cannot be found. Therefore actin in the electron dense area may not be polymerized to form F-actin, but exists as G-actin.

There is no de novo synthesis of actin prior to swelling in the germination process, and spores can swell under conditions that prohibit de novo protein synthesis. Reconstruction of a functional actin cytoskeleton in swollen spores likely relies upon existing actin molecules that compose actin rods as well as G-actin aggregates.

TEM of 2 days old spores quickly frozen by dipping into liquid propane and freeze-subutituted is shown in (**A**). Arrows show actin rods in the cytoplasm and the nucleus. A cross-sections of a rod is shown in (**B**). Enlarged cross section of an actin tubule is shown in (**C**). G-actin aggregates surrounded by lipid droplets is shown in (**D**). Scale bars: 1 μm (**A**), 20 nm (**B**), 10 nm (**C**), 200 nm (**D**).

Contributors

Masazumi Sameshima*, Integrated Imaging Research Support, Villa Royal Hirakawa 103, 1-7-5 Hirakawa-cho, Chiyoda-ku, Tokyo 102-0093, Japan
*E-mail: samesima@beige.ocn.ne.jp

References

27. Sameshima M, Kishi Y, Osumi M, Mahadeo D, Cotter D (2000) Novel actin cytoskeleton: actin tubules. Cell Struct Funct 25:291–295
28. Sameshima, M (2012) Stabilization of dormant spores depends on the actin cytoskeleton in the cellular slime mold. Plant Morphol 24:65–71
29. Sameshima M, Kishi Y, Osumi M, Minamikawa-Tachino R, Mahadeo D, Cotter D (2001) The formation of actin rods composed of actin tubules in *Dictyostelium discoideum* spores. J Struct Biol 136:7–19

6 Cytoskeletons

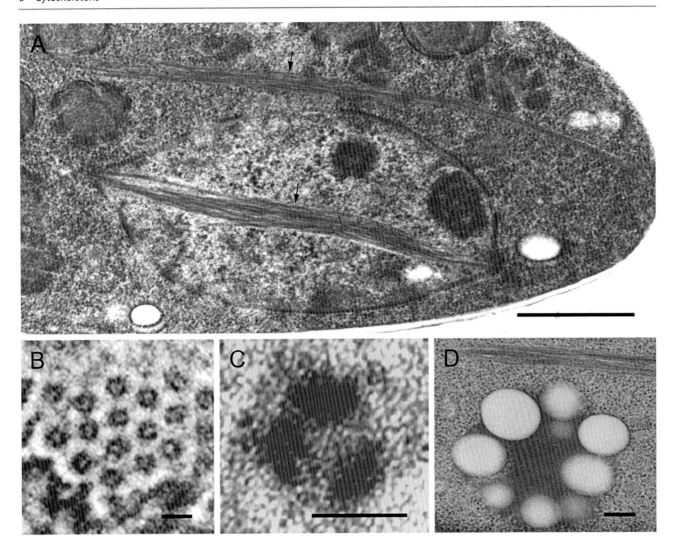

Plate 6.13

Actin-microtubule interaction during preprophase band formation in onion root tips visualized by immuno-fluorescence microscopy

Actin filaments and microtubules (MTs) are two major cyto-skeleton components and are known to play important roles in eukaryotic cellular and morphogenetic processes. Preprophase band (PPB) formation is one morphogenetic process in which actin-MT interactions are thought to change during development. The PPB is observed as an array of cortical MTs (PPB MT) located in the preprophasic cell cortex of plant cells. It originates as a broad band of MTs during the G2 phase and matures into a narrow band during prophase. The actin depolymerizing agent cytochalasin affects MT band narrowing in the PPB and treatment causes a shift of the newly formed cell walls to a more central region in the unequal division for guard mother cell formation [30]. Actins are reported to be in the PPB in early stages of PPB development, but they are excluded from the PPB in late prophase, and this actin-depleted zone (ADZ) remains even after the disappearance of the MT band. As the mature PPB predicts the site where the cell plate joins the parent walls, the ADZ is a candidate of negative division-site memory [31].

This figure shows onion seedling root tip cells (*Allium cepa* L. cv. Highgold Nigou) triple labeled for actin (red), tubulin (green) and DNA (Hoechst 33258 staining, blue) during PPB development. In G2 phase (Interphase), a broad PPB MT is formed. The PPB MT narrows during prophase (Prophase 1), and MTs are tightly packed in the PPB (Prophase 2). Cortical actins also appear in the PPB region in these phases (Interphase, Prophase 1 and Prophase 2). However, actin disappears from the PPB region (Prophase 3) to form ADZ (arrowheads) when the PPB matures. Actin filaments in plant cells are fragmented by cytochalasin treatment. When cytochalasin D is applied to late prophase cells, PPB MT widening occurs and actins stay in the broadened PPB instead of moving out from the PPB (Cytochalasin D). Scale bar: 10 μm.

References

30. Mineyuki Y, Palevitz PA (1990) Relationship between preprophase band organization, F-actin and the division site in *Allium*. J Cell Sci 97:283–295
31. Mineyuki Y (1999) The preprophase band of microtubules: its function as a cytokinetic apparatus in higher plants. Int Rev Cytol 187:1–49

Contributors

Miyuki Takeuchi[1], Yoshinobu Mineyuki[2]*, [1]Department of Biomaterial Sciences, Graduate School of Agricultural and Life Sciences, The University of Tokyo, 1-1-1 Yayoi, Bunkyo-ku, Tokyo 113-8657, Japan, [2]Department of Life Science, Graduate School of Life Science, University of Hyogo, 2167 Shosha, Himeji, Hyogo 671-2280, Japan
*E-mail: mineyuki@sci.u-hyogo.ac.jp

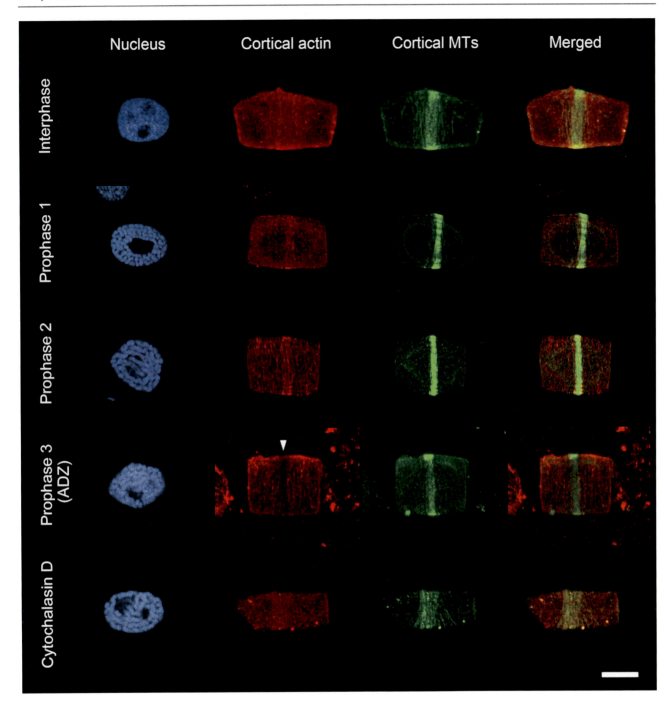

Plate 6.14

Microtubules direct the layered structure of angiosperm shoot apical meristems (SAMs)

Cortical microtubules (cMTs) are situated just beneath the plasma membrane in interphase plant cells. In general, their orientation coincides with that of cellulose microfibrils in the adjacent cell wall, because a cellulose synthesizing complex moves along the cMT while leaving newly formed cellulose fibrils behind [32]. As cellulose microfibrils mechanically reinforce the cell wall, especially parallel to their direction, the cell expands perpendicularly to the direction of cMTs. As such, cMTs determine the direction of cell growth by controlling the mechanical properties of the cell wall.

One salient feature of angiosperm SAMs is that several layers of cells ("tunica") cover the internal cell mass ("corpus"); such a tissue configuration is known as the tunica-corpus theory, first proposed by Schmidt (1924) [33], which is not found in SAMs of ferns nor most gymnosperms. The layered structure is genetically important, because initials (or stem cells) exist independently in each cell layer (often called L1, L2, L3), and give a chimeric nature to the angiosperm shoot. If a mutation occurs in the initials of one layer, or if cells with a different genotype are brought into the initials by grafting, a chimeric body is formed which is perpetually maintained in the shoot, its branches, and the individuals vegetatively propagated from them.

Photographs show anti-tubulin immunofluorescence of a median longitudinal section of a SAM in the vegetative phase of a periwinkle (*Vinca major* L.) plant (adapted from [34]). A shoot tip was fixed with 9 % formaldehyde/10 % dimethylsulfoxide in phosphate buffer, sectioned at 7 µm by cryomicrotome, treated with a Nonidet P-40 detergent solution and a mixed solution of cellulose and pectolyase, then stained with a mixture of mouse monoclonal anti-chicken-α-tubulin and mouse monoclonal anti-chicken-β-tubulin, followed by FITC-labeled goat anti-mouse IgG treatment. The resulting section was observed by blue light excitation with a conventional epifluorescence microscope.

Image of whole SAM is shown (**C**). In tunica, vertically arranged cMTs (fine threads in the cells) are observable, coinciding with lateral cell growth within the cell layer (**A**). In corpus, cMT directions are mixed, corresponding to cell growth in all directions resulting in increments of cell mass volume (**B**). A region beneath an apical dome extending to stem tissues, the rib meristem, where longitudinal files of cells are developed, shows transversely arranged cMTs, again coinciding with vertical cell growth (**D**). Thus cMTs provide a basis for characteristic structure of angiosperm SAM. A, B and D are magnified views of the portions of C with white stars, circles and triangles indicating the corresponding positions, respectively. Scale bar: 10 µm.

Contributors

Shuichi Sakaguchi*, Course of Biological Sciences, Faculty of Science, Nara Women's University, Kitauoya-nishimachi, Nara 630-8506, Japan
*E-mail: guchi@cc.nara-wu.ac.jp

References

32. Paredez AR, Somerville CR, Ehrhardt DW (2006) Visualization of cellulose synthase demonstrates functional association with microtubules. Science 312:1491–1495
33. Schmidt A (1924) Histologische Studien an phanerogamen Vegetationspunkten. Bot Arch 8:345–404
34. Sakaguchi S, Hogetsu T, Hara N (1988) Arrangement of cortical microtubules in the shoot apex of *Vinca major* L. Observations by immunofluorescence microscopy. Planta 175:403–411

6 Cytoskeletons

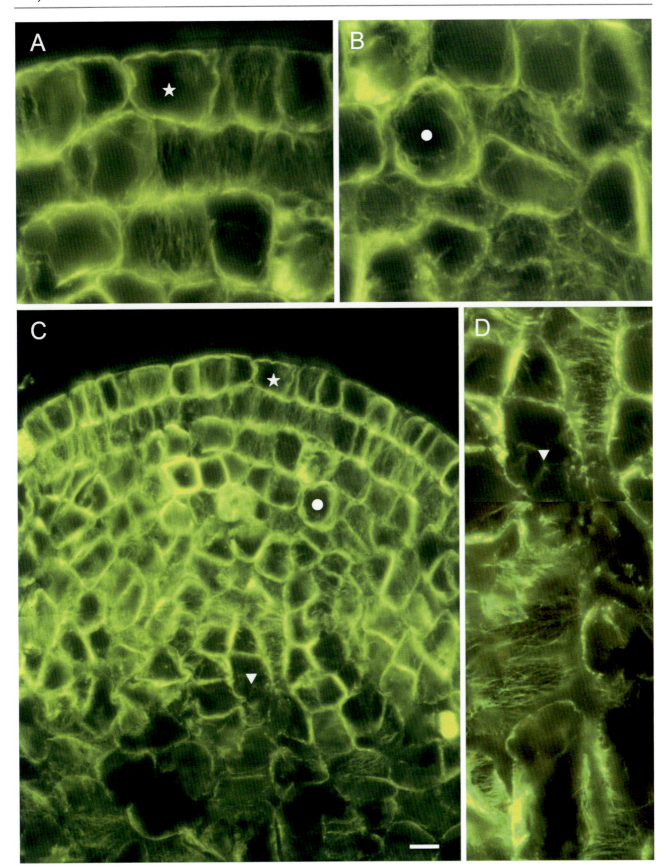

Chapter References

1. Mineyuki Y (2007) Plant microtubule studies: past and present. J Plant Res 120:45–51
2. Mineyuki Y, Iida H, Anraku Y (1994) Loss of microtubules in the interphase cells of onion (*Allium cepa* L.) root tips from the cell cortex and their appearance in the cytoplasm after treatment with cycloheximide. Plant Physiol 104:281–284
3. Nogami A, Suzaki T, Shigenaka Y, Nagahama Y, Mineyuki Y (1996) Effects of cycloheximide on preprophase bands and prophase spindles in onion (*Allium cepa* L.) root tip cells. Protoplasma 192:109–121
4. Murata T, Sonobe S, Baskin TI, Hyodo S, Hasezawa S, Nagata T, Horio T, Hasebe M (2005) Microtubule-dependent microtubule nucleation based on recruitment of γ-tubulin in higher plants. Nat Cell Biol 7:961–968
5. Petry S, Groen AC, Ishihara K, Mitchison TJ, Vale RD (2013) Branching microtubule nucleation in Xenopus egg extracts mediated by augmin and TPX2. Cell 152:768–777
6. Kamasaki T, O'Toole E, Kita S, Osumi M, Usukura J, McIntosh JR, Goshima G (2013) Augmin-dependent microtubule nucleation at microtubule walls in the spindle. J Cell Biol 202:25–33
7. Karahara I, Kang BH (2014) High-pressure freezing and low-temperature processing of plant tissue samples for electron microscopy. Methods Mol Biol 1080:147–157
8. Murata T, Karahara I, Kozuka T, Giddings TH Jr, Staehelin LA, Mineyuki Y (2002) Improved method for visualizing coated pits, microfilaments, and microtubules in cryofixed and freeze-substituted plant cells. J Electron Microsc 51:133–136
9. Karahara I, Suda J, Tahara H, Yokota E, Shimmen T, Misaki K, Yonemura S, Staehelin A, Mineyuki Y (2009) The preprophase band is a localized center of clathrin-mediated endocytosis in late prophase cells of the onion cotyledon epidermis. Plant J 57:819–831
10. Karahara I, Suda J, Staehelin LA, Mineyuki Y (2009) Quantitative analysis of vesicles in the preprophase band by electron tomography. Cytologia 74:113–114
11. Mineyuki Y, Suda J, Karahara I (2004) Electron tomography. Plant Morphol 16:21–30 (Japanese)
12. Mineyuki Y (2013) Electron tomography and structure of plant cell framework. In: IIRS (eds) Structure and function of life analyzed by 3D imaging. Asakura Publishing Co. Ltd., Tokyo, pp 51–60. ISBN 978-4-254-17157-0 C3045 (Japanese)
13. Shimamura M, Brown RC, Lemmon BE, Akashi T, Mizuno K, Nishihara N, Tomizawa KI, Yoshimoto K, Deguchi H, Hosoya H, Horio T, Mineyuki Y (2004) γ-Tubulin in basal land plants: characterization, localization, and implication in the evolution of acentriolar microtubule organizing centers. Plant Cell 16:45–59
14. Shimamura M (2004) Monoplastidic cell in lower land plants. Plant Morphol 16:83–92
15. Motomura T, Nagasato C, Kimura K (2010) Cytoplasmic inheritance of organelles in brown algae. J Plant Res 123:185–192
16. Nagasato C, Motomura T, Ichimura T (1998) Selective disappearance of maternal centrioles after fertilization in the anisogamous brown alga *Cutleria cylindrica* (Cutleriales, Phaeophyceae): paternal inheritance of centrioles is universal in the brown alga. Phycol Res 46:191–198
17. Motomura T (1994) Electron and immunofluorescence microscopy on the fertilization of *Fucus distichus* (Fucales, Phaeophyceae). Protoplasma 178:97–110
18. Nagasato C, Motomura T (2002) Ultrastructural study on mitosis and cytokinesis in *Scytosiphon lomentaria* zygotes (Scytosiphonales, Phaeophyceae) by freeze-substitution. Protoplasma 219:140–149
19. Yubuki N, Leander BS (2011) Reconciling the bizarre inheritance of microtubules in complex (euglenid) microeukaryotes. Protoplasma 249:859–869
20. Jaspersen SL, Winey M (2004) The budding yeast spindle pole body: structure, duplication, and function. Annu Rev Cell Dev Biol 20:1–28
21. Hirata A (2010) Meiosis I in *Saccharomyces cerevisiae* by rapid-freeze electron microscopy. Cytologia 75:221–222 (Technical note)
22. Tanaka I, Kitazume C, Ito M (1987) The isolation and culture of lily pollen protoplasts. Plant Sci 50:205–211
23. Tanaka I, Wakabayashi T (1992) Organization of the actin and microtubule cytoskeleton preceding pollen germination: an analysis using cultured pollen protoplasts of *Lilium longiflorum*. Planta 186:473–482
24. Riedl J, Crevenna AH, Kessenbrock K, Yu JH, Neukirchen D, Bista M, Bradke F, Jenne D, Holak TA, Werb Z, Sixt M, Wedlich-Sordner R (2008) Lifeact: a versatile marker to visualize F-actin. Nat Methods 5:605–607
25. Era A, Tominaga M, Ebine K, Awai C, Saito C, Ishizaki K, Yamato KT, Kohchi T, Nakano A, Ueda T (2009) Application of lifeact reveals F-actin dynamics in *Arabidopsis thaliana* and the liverwort, *Marchantia polymorpha*. Plant Cell Physiol 50:1041–1048
26. Era A, Kutsuna N, Higaki T, Hasezawa S, Nakano A, Ueda T (2013) Microtubule stability affects the unique motility of F-actin in *Marchantia polymorpha*. J Plant Res 126:113–119
27. Sameshima M, Kishi Y, Osumi M, Mahadeo D, Cotter D (2000) Novel actin cytoskeleton: actin tubules. Cell Struct Funct 25:291–295
28. Sameshima M (2012) Stabilization of dormant spores depends on the actin cytoskeleton in the cellular slime mold. Plant Morphol 24:65–71
29. Sameshima M, Kishi Y, Osumi M, Minamikawa-Tachino R, Mahadeo D, Cotter D (2001) The formation of actin rods composed of actin tubules in *Dictyostelium discoideum* spores. J Struct Biol 136:7–19
30. Mineyuki Y, Palevitz PA (1990) Relationship between preprophase band organization, F-actin and the division site in *Allium*. J Cell Sci 97:283–295
31. Mineyuki Y (1999) The preprophase band of microtubules: its function as a cytokinetic apparatus in higher plants. Int Rev Cytol 187:1–49
32. Paredez AR, Somerville CR, Ehrhardt DW (2006) Visualization of cellulose synthase demonstrates functional association with microtubules. Science 312:1491–1495
33. Schmidt A (1924) Histologische Studien an phanerogamen Vegetationspunkten. Bot Arch 8:345–404
34. Sakaguchi S, Hogetsu T, Hara N (1988) Arrangement of cortical microtubules in the shoot apex of *Vinca major* L. Observations by immunofluorescence microscopy. Planta 175:403–411

Cell Walls

7

Tetsuko Noguchi

The cell wall is an extracellular structure that encloses each cell in land plants, algae, and fungi. It is more rigid, thicker, and stronger than the extracellular matrix produced by animal cells. Typical cell walls in land plants are composed predominantly of polysaccharides, mainly cellulose microfibrils, which are embedded in a matrix of pectin and hemicellulose. Cellulose is synthesized by cellulose-synthesizing complexes localized on the plasma membrane, which spin out crystalline cellulose microfibrils. In contrast, hemicelluloses and pectins are synthesized in Golgi bodies, carried to the plasma membrane by vesicles, and secreted into the cell wall, where hemicelluloses cross-link with cellulose microfibrils and pectins fill the space in between to form the three dimensional dynamic structure of the cell wall.

The plant cell wall plays important roles in controlling cell differentiation. The thin wall of a young cell, called the primary cell wall, contains many enzymes, including not only glycosyl hydrolases, which simply disassemble cell wall polysaccharides, but also cell-wall modifying enzymes such as expansin, endoxyloglucan transferase/hydrolase, and pectinmethyl esterase, which are implicated in the remodeling of cell wall architecture, and are required for regulation of cell elongation and differentiation. After cell elongation ceases, the cell wall is generally thickened by formation of a secondary cell wall. Since cell walls connect cells to form tissues and provide the tissues with mechanical strength, they determine the growth and development of the entire plant body. Within the secondary cell wall is deposited lignin, which is essential in allowing land plants to strengthen vascular tissues. In addition, the cell wall works as an external sensor in defending against and responding to environmental stresses.

Cellulose is also the main component of algae and fungi cell walls. Compared to land plants, the process of cell wall formation can be more easily observed and analyzed in these organisms. Unicellular algae demonstrate a variety of shapes, which are determined by their primary cell walls according to the specialized accumulation of cell wall materials.

This chapter illustrates a variety of cell wall formations in fungal, algal, and plant cells. M. Osumi presents the excreted ribbon-like fibrillar network of glucan framing the cell wall in a reverting protoplast of fission yeast, *Schizosaccharomyces pombe*, by low-voltage scanning electron microscopy. M. Yamamoto and S. Kawano show cell wall formation during autosporulation in the green alga *Chlorella vulgaris*, in which wall synthesis begins before successive protoplast division, using fluorescent microscopy as well as a rapid-freeze-fixation method and transmission electron microscopy. M. Yamamoto et al. also demonstrate four generations of cell walls in the green alga *Marvania geminata* by fluorescent microscopy and field emission scanning electron microscopy. S. Sekida and K. Okuda demonstrate the formation of a complex armor-like covering called the amphiesma, which is composed of a series of flattened vesicles containing cellulose plates, in the dinoflagellate *Scrippsiella hexapraecingula* using thin-section and freeze-fracture methods and electron microscopy, as well as fluorescent microscopy. T. Noguchi detects pectin in the primary cell wall of developing daughter semi-cells of *Micrasterias* by fluorescent microscopy. In addition, she shows cellulose-synthesizing rosettes in the plasma membranes of secondary cell walls forming *Micrasterias* and a *Closterium* zygote using freeze-fracture methods and electron microscopy. R. Yokoyama et al. demonstrate the immunological localization of two typical cell wall polysaccharides, pectin and b-1,3/1,4 mixed linked glucan, individually in two angiosperms, *Arabidopsis thaliana* and *Oryza sativa*, using fluorescent and electron microscopy.

The cell wall encloses each cell while at the same time enabling the transfer of solutes and signaling between cells via plasmodesmata. Casparian strips are a chemically modified region of the cell wall in the endodermis of vascular plant roots. The Casparian strip acts as a barrier that is thought to be crucial for selective nutrient uptake and exclusion of pathogens.

Y. Hayashi shows the distribution of plasmodesmata in the cell wall of cotyledons by osmium tetraoxide-potassium ferricyanide staining and electron microscopy. Y. Honma and I. Karahara successfully show an isolated Casparian strip from a pea root and its meshwork structure by fluorescent microscopy.

T. Noguchi (✉)
Course of Biological Sciences, Faculty of Science, Nara Women's University, Kitauoya-nishimachi, Nara 630-8506, Japan
e-mail: noguchi@cc.nara-wu.ac.jp

T. Noguchi et al. (eds.), *Atlas of Plant Cell Structure*,
DOI 10.1007/978-4-431-54941-3_7, © Springer Japan 2014

Plate 7.1

Ribbon-like fibrillar network of glucan in reverting *Schizosaccharomyces pombe* protoplast

Yeast, a typical fungus, has been used as a model system for various basic and applied fields of life science, medicine and biotechnology. Yeast sub-cellular structure is fundamentally the same as higher animal and plant cell structure. The cell wall is the sole yeast structure which is not found in animal cells. The cell wall is situated on the outer surface of the cell and is important for maintaining the genetically determined cell shape. The cell wall plays an important role in the transportation of materials into and out of the cell. The cell wall is the first line of defense of the host to a fungal infection from the standpoint of pathology, and in chemotherapy it is a primary target of antifungal agents. Analysis of the mechanisms of cell wall formation is therefore important in both basic and applied biology [1].

The ultrastructual image of the yeast cell was first described by Agar and Douglas in 1957 based on imaging of thin sections fixed with potassium permanganate ($KMnO_4$). Chemical fixation and embedding techniques present many problems in the preservation of membranous structures within yeast cells. Direct osmium tetroxide (OsO_4)-fixation results in poor preservation for all structures except protoplasts. Until the 1970s, only $KMnO_4$ was known to preserve yeast membrane architecture and show the membrane lipid bilayer with high electron density. In early ultrastructural studies of yeast, cells were fixed with $KMnO_4$ and embedded with methacrylate because OsO_4 and epoxy resin were found to not penetrate the thick cell walls. The first phase of yeast ultrastructural studies were in the 1960s, and organelles, especially mitochondria, were studied by single fixation with $KMnO_4$ or double fixation with GA-$KMnO_4$. In the 1970s GA-OsO_4 fixation was generally used following digestion of the cell wall by Zymolase (β-1, 3-glucanase). This method allows us to study yeast cells with rigid cell walls, which had been the principal obstacle to earlier physiological and morphological studies. In the 1980s the rapid freeze substitution technique was difficult to apply to yeast cells, but was modified by application of the 'sandwich method.' Although this technique is quite complicated, it allowed study of yeast cell ultrastructure at the same resolution level as is possible for other biological specimens [1].

This image is a low voltage scanning electron microscope (LVSEM) image of the ribbon-like fibrillar network of glucan framing the cell wall 5 h after it was regenerated from a reverting protoplast in fission yeast, *S. pombe*. In approximately 7 h these cells will revert to the original cylindrical rod shape covered by galactomannan. This is the first image of a reverting protoplast photographed by LVSEM [2, 3].

Specimens were fixed with 2 % GA for 2 h, postfixed in 2 % OsO_4 for 1 h or with 2 % ruthenium tetroxide for 7 min, and prepared for SEM in the usual manner. Specimens were slightly coated with a 2 nm layer of platinum-carbon at 2×10^7 Pa and 10 °C in a Balzers 500 K with an electron gun and examined with the in-lens FESEM, Hitachi S-900, at 1–3 kV [3, 4]. Scale bar: 1 µm

Contributors

Masako Osumi*, Integrated Imaging Research Support, Villa Royal Hirakawa 103, 1-7-5-103 Hirakawa-Cho, Ciyoda-ku, Tokyo l02-0093, Japan
*E-mail: osumi@fc.jwu.ac.jp

References

1. Osumi M, (1998) The ultrastructure of yeast: cell wall structure and formation. Micron 29:207–233
2. Osumi M, Yamada N, Kobori H, Naito N, Baba M, Nagatani T (1989) Cell wall formation in regenerating protoplasts of *Schlzosaccharomyces pombe*. J Electron Microsc 38:457–468
3. Osumi M (2012) Visualization of yeast cells by electron microscopy. J Electron Microsc 61:343–365
4. Osumi M, Yamada N, Yaguchi H, Kobori H, Nagatani T, Sato M (1995) Ultrahigh-resolution low-voltage SEM reveals ultrastructure of the glucan network formation fission yeast protoplast. J Electron Microsc 44:198–206

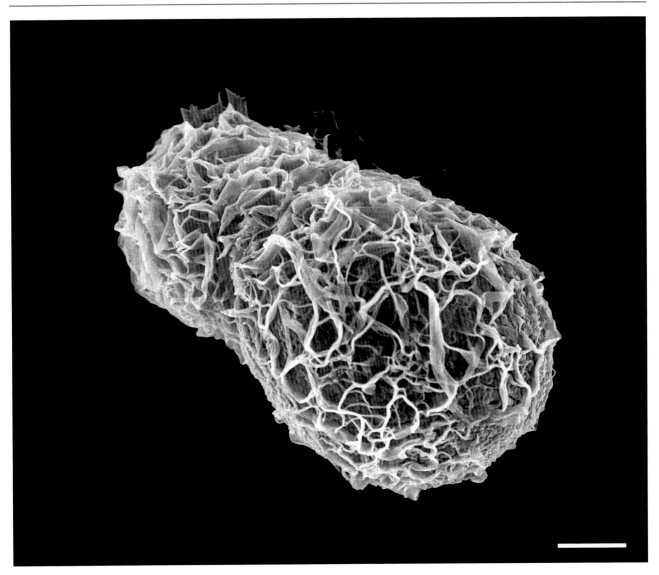

Plate 7.2

Mother and daughter cell walls during autosporulation in the green alga *Chlorella vulgaris*

Chlorella propagates by autosporulation, which is the formation of daughter cells with their own cell walls enveloping their protoplast. The structural changes which occur with growth of premature daughter cell walls during the mother cell-division phases in *C. vulgaris* were examined using the cell wall-specific fluorescent dye Fluostain I (**A**) and electron microscopy (**B**). Two clearly distinguishable stages of daughter cell wall synthesis are suggested by fluorescence observations: moderate synthesis occurs during the mother cell growth process and rapid synthesis occurs during the mother cell division phase. Daughter cell wall synthesis occurs over the cell surface in the early stage of the mother cell growth process, as indicated by electron microscopy studies. The newly synthesized daughter cell wall gradually becomes thick. After the successive second protoplast division, each daughter cell matured to a round shape. During the process of autospore maturation, the daughter cell wall (arrowhead) rapidly increased in thickness and reached substantially the thickness of the mother cell wall (double arrowhead) before hatching [5, 6].

For fluorescent microscopy of cell wall components with Fluostain I, cultured cells were stained with 0.001 % Fluostain I in PBS buffer. Stained samples were observed under ultraviolet excitation of Fluostain I using a fluorescence microscope. Stained cells were imaged by a chilled CCD camera system. For electron microscopy, cultured cells were fixed using the rapid freeze fixation method, followed by freeze substitution with 2.5 % glutaraldehyde in dry acetone. Samples were subsequently transferred to 2 % OsO_4 in dry acetone at 40 °C for 4 h. Ultra-thin sections were stained with 3 % uranyl acetate for 2 h at room temperature, with lead citrate for 10 min at room temperature, and finally examined with a transmission electron microscope at 100 kV. Scale bars: 5 μm (**A**), 500 nm (**B**). This figure is adapted from [5, 6].

References

5. Yamamoto M, Fujishita M, Hirata A, Kawano S (2004) Regeneration and maturation of daughter cell walls in the autospore-forming green alga *Chlorella vulgaris* (Chlorophyta, Trebouxiophyceae). J Plant Res 117:257–264
6. Yamamoto M, Kawano S (2004) Daughter cell wall synthesis in autorpore-forming green alga, *Chlorella vulgaris* (IAM-C-536). Cytologia 69–3:i–ii

Contributors

Maki Yamamoto[1]*, Shigeyuki Kawano[2], [1]Institute of Natural Sciences, Senshu University, 2-1-1 Higashimita, Tama, Kawasaki, Kanagawa 214-8580, Japan, [2]Department of Integrated Biosciences, Graduate School of Frontier Sciences, The University of Tokyo, Bldg. FSB-601, 5-1-5 Kashiwanoha, Kashiwa, Chiba 277-8562, Japan
*E-mail: yamamoto@isc.senshu-u.ac.jp

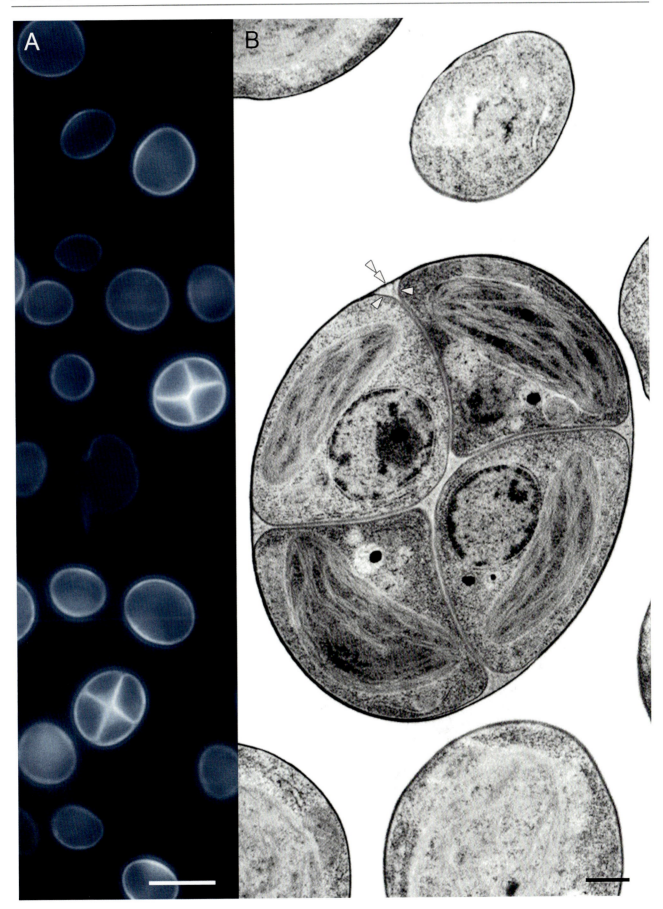

Plate 7.3

Great-grandmother, grandmother, mother, and daughter cell walls during budding in the green alga *Marvania geminata*

Marvania geminata propagates by budding. To examine structural changes in cell wall structure during the *M. geminata* cell cycle, we used Fluostain I, a cell wall-specific fluorescent dye (**A–C**), and field emission scanning electron microscopy (**D**). Cells in early growth phase were spherical (**A**), and Fluostain I fluorescence was observed at cell edges and in the upper and lower right-hand side of the image. Cells expanded by budding from the position of strong Fluostain I fluorescence to the right side of the image (**B**). During protoplast division, a belt of Fluostain I fluorescence was ascertained at the budding site (i.e., at the plane of division), which was the position of strong fluorescence during the growth phase. Subsequently, the two daughter cells separated (**C**) [7]. The daughter cell on the left side of the image is covered by great-grand mother (arrow), grandmother (triple arrowheads), mother (double arrowheads) and daughter cell walls (arrowhead); the daughter cell on the right side is covered by mother (double arrowheads) and daughter cell walls (arrowhead) (**D**). The two daughter cells were substantially asymmetrical from the stand of cell wall formation. Daughter cells entered the next cell cycle still retaining the mother and grandmother cell wall, causing the uncovered site of the cell to bud outward [8].

For fluorescence microscopy of cell wall components, cultured cells were stained with 0.001 % Fluostain I in PBS buffer. Stained samples were observed under ultraviolet excitation of Fluostain I using a fluorescence microscope. Samples were imaged with a chilled CCD camera system. For field emission scanning electron microscopy, cells were fixed with 2.5 % glutaraldehyde in 0.1 M phosphate buffer (pH 7.4) for 2 h, washed three times (20 min × 3) and post-fixed with 1 % OsO_4 for 1 h at room temperature. To take away the mucilage around the *Marvania* cell wall, cells were treated with 5 % sodium hypochlorite for 10 min before postfixation. Cells were next washed two to six times for 15 min in Milli-Q water (Millipore, Billerica). Specimens were dehydrated with a graded ethanol series (30–100 %) followed by isoamyl acetate (50 % in ethanol and 100 %), and dried with a Hitachi critical point dryer (HCP-2; Hitachin). Specimens were coated with osmium using a Neoc osmium coater (Meiwafosis Co., Ltd), and imaged using an S-5000 field emission scanning electron microscope (Hitachi). Scale bars: 5 μm (**C**), 1 μm (**D**). This figure is adapted from [7, 8].

Contributors

Maki Yamamoto[1]*, Satomi Owari[2], Shigeyuki Kawano[3], [1]Institute of Natural Sciences, Senshu University, 2-1-1 Higashimita, Tama, Kawasaki, Kanagawa 214-8580, Japan, [2]Neo-Morgan Laboratory Incorporated Research & Development, Biotechnology Research Center, 907 Nogawa, Miyamae-ku, Kawasaki, Kanagawa 216-0001, Japan, [3]Department of Integrated Biosciences, Graduate School of Frontier Sciences, The University of Tokyo, Bldg. FSB-601, 5-1-5 Kashiwanoha, Kashiwa, Chiba 277-8562, Japan
*E-mail: yamamoto@isc.senshu-u.ac.jp

References

7. Yamamoto M, Nishikawa T, Kajitani H, Kawano S (2007) Patterns of asexual reproduction in *Nannochloris bacillaris* and *Marvania geminata* (Chlorophyta, Trebouxiophyceae). Planta 226:917–927
8. Yamazaki T, Owari S, Ota H, Sumiya N, Yamamoto M, Watanabe K, Nagumo T, Miyamura S, Kawano, S (2013) Localoization and evolution of septins in algae. Plant J 74:605–614

7 Cell Walls

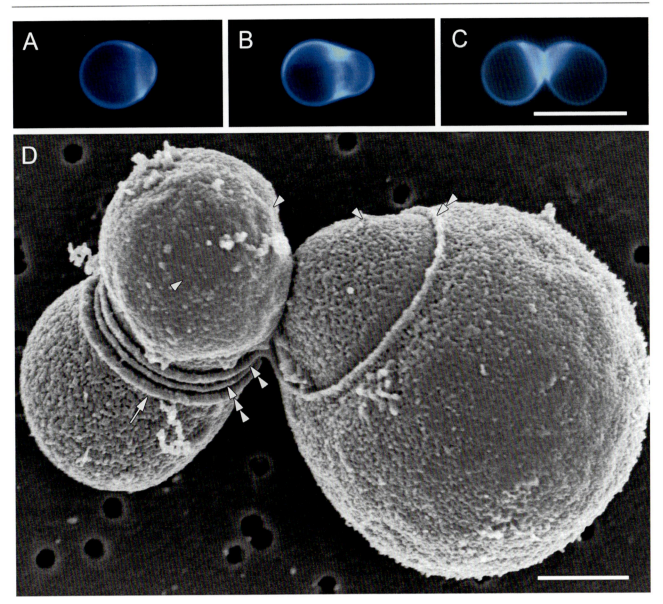

Plate 7.4

Formation of amphiesmal vesicles and thecal plates in the dinoflagellate *Scrippsiella hexapraecingula*

Scrippsiella hexapraecingula is an armored, peridinioid dinoflagellate that has a simple, asexual life cycle in which cultured cells alternate diurnally between motile and non-motile forms. Motile cells transform into nonmotile cells through ecdysis. During the nonmotile phase, one or two daughter cells which develop new amphiesmal vesicles are produced inside the pellicle [9]. A cross section of a nonmotile cell fixed 6 h after ecdysis shows a thickened pellicle at the outside of the plasma membrane (pm) and developing amphiesmal vesicles (asterisks) with an outer amphiesmal vesicle membrane (arrow) and an inner amphiesmal vesicle membrane (double arrowheads) (**A**). Amphiesmal vesicles are empty but arranged in the same pattern as the developing thecal plates. Freeze-fracture images of nonmotile cells fixed 2 h (**B**) and 9 h (**C**) after ecdysis show the protoplasmic faces of outer amphiesmal vesicle membranes (oam PF) and exoplasmic faces of inner amphiesmal vesicle membranes (iam EF) and the plasma membrane (pm EF). Early amphiesmal vesicle-like patches developed within territories enclosed by broken lines in a cell 2 h after ecdysis (**B**), whereas adjacent almost complete amphiesmal vesicles came in contact with each other to form sutures (arrowheads) along the boundaries of territories in a cell 9 h after ecdysis (**C**) (adapted from [9]). Thecal plates are produced in the amphiesmal vesicles of motile cells [10], and therefore the thecal plate pattern is determined at the time of development of amphiesmal vesicles in nonmotile cells [11]. Motile cells were fixed and stained 5, 10, 15 and 30 min after they escaped from pellicles of nonmotile cells and began to swim (**D**). Incipient thecal plates appear first as groups of granular materials (blue) which increase in number, such that thin sheet-like thecal plates develop. Plate materials spread continuously within amphiesmal vesicles, and finally individual thecal plates become sufficiently close to each other to be arranged in a pattern specific to this species.

For thin sectioning, nonmotile cells were fixed by a cold block-freezing method. Freeze substitution with a 4:1 (v/v) mixture of acetone and methanol containing 2 % OsO_4 was carried out at $-80\ °C$ for 30 h. The sample was warmed gradually to room temperature, rinsed with ethanol, and embedded in LR White resin. For freeze fracture, nonmotile cells were frozen rapidly in nitrogen slush. Freeze fracture and metal shadowing were performed with a Baltec BAF 060 apparatus (Baltec Inc.) at $-106\ °C$ and 1×10^{-6} mbar. Specimens were shadowed unidirectionally at an angle of $60°$ with platinum-carbon and subsequently coated with

carbon to make replicas. For fluorescence microscopy, motile cells were fixed with seawater containing 1.2 % glutaraldehyde and 0.4 % OsO_4 at 4 °C for 10 min, rinsed with seawater, and stained with seawater containing 0.1 % Calcofluor White M2R (to stain thecal plate materials) and 1 μg/mL ethidium bromide (to stain nuclei). Samples were imaged with an epifluorescence microscope using ultraviolet excitation. Scale bars: 200 nm (**A**), 1 μm (**B**), 500 nm (**C**), 50 μm (**D**)

Contributors

Satoko Sekida, Kazuo Okuda*, Graduate School of Kuroshio Science, Kochi University, 2-5-1 Akebono-cho, Kochi 780-8520, Japan
*E-mail: okuda@kochi-u.ac.jp

References

9. Sekida S, Horiguchi T, Okuda K (2001) Development of the cell covering in the dinoflagellate *Scrippsiella hexapraecingula* (Peridiniales, Dinophyceae). Phycological Res 49:163–176
10. Sekida S, Horiguchi T, Okuda K (2004) Development of thecal plates and pellicle in the dinoflagellate *Scrippsiella hexapraecingula* (Peridiniales, Dinophyceae) elucidated by changes in stainability of the associated membranes. Eur J Phycol 39:105–114
11. Sekida S, Takahira M, Horiguchi T, Okuda K (2012) Effects of high pressure in the armored dinoflagellate *Scrippsiella hexapraecingula* (Peridiniales, Dinophyceae): changes in thecal plate pattern and microtubule assembly. J Phycol 48:163–173

7 Cell Walls

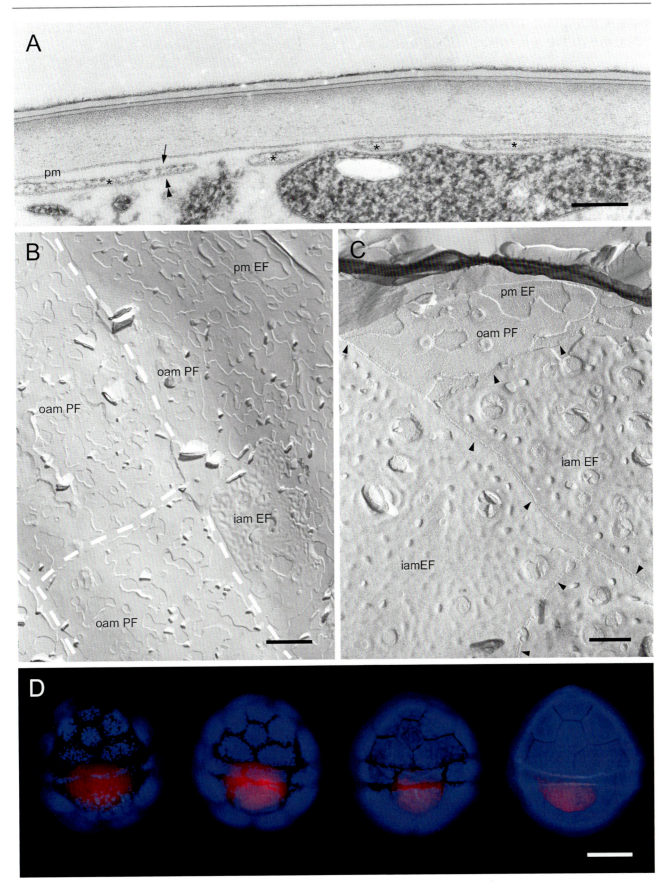

Plate 7.5

The elaborate shape of *Micrasterias* is formed by a primary cell wall containing pectin

The unicellular green algae *Micrasterias* is composed of two semi-cells joined at a deep median construction. Each semi-cell has an elaborate outer shape composed of a number of lobes. The nucleus lies in the central isthmus defined by the median construction. After mitosis, the cell is divided into two cells by a septum across the isthmus. Each daughter cell begins to grow as a hemispherical bulge, which later develops 3, 5, and then 9 lobes. The development of this elaborate cell has attracted the attention of many cell biologists as a model system for cell morphogenesis research.

The primary cell wall, which contains pectic substances, begins to grow after septum formation (upper cell) and continues until the daughter hemi-cell develops fully (lower cell). The main components of the primary cell wall are synthesized in Golgi bodies and carried by special vesicles which have a dark electron dense core and large vesicles when imaged by electron microscopy [12]. The primary wall is decomposed after the formation of the cellulosic secondary wall as shown by the weak fluorescence of the cell wall of mother semi-cells, which has also been confirmed by electron microscopy. Daughter cell shape is determined by the shape of the developed primary wall, which is induced by precocious differentiation of the wall at the sinus [13].

This figure is a fluorescent micrograph of growing *Micrasterias* sp. immunolabeled with an antibody against pectin purified from the green alga *Botryococcus braunii*.

Cells were fixed with 3 % paraformmaldehyde, rinsed, treated with a monoclonal anti-pectin antibody, then treated with FITC-labeled anti-mouse IgG. Scale bar: 10 μm.

Contributors

Tetsuko Noguchi*, Course of Biological Sciences, Faculty of Science, Nara Women's University, Kitauoya-nishimachi, Nara 630-8506, Japan
*E-mail: noguchi@cc.nara-wu.ac.jp

References

12. Ueda K, Noguchi T (1976) Transformation of the Golgi apparatus in the cell cycle of a green alga, *Micrasterias americana.* Protoplasma 87:145–162
13. Ueda K, Yoshioka S (1976) Cell wall development of *Micrasterias americana*, especially in isotonic and hypertonic solutions. J Cell Sci 21:617–631

Plate 7.6

Cellulose-synthesizing rosettes in the green algae *Micrasterias* and *Closterium*

The secondary cell wall of *Micraterias* is composed of cellulose microfibrils arranged in a parallel orientation that form layers of crossed microfibrils (right region in (**A**) adapted from [14]). The synthesis of the cellulosic secondary cell wall starts 6 h after daughter semi-cells are well developed (**B**, yellow; cellulosic secondary cell wall, red; chloroplast).

Freeze fracture electron microscopy allows visualization of the membrane interior, in which the fracture plane often passes through the hydrophobic interior of membrane lipid bilayers. In fully grown daughter semi-cells, the P-fracture face of the plasma membrane is composed of a hexagonal array of cellulose-synthesizing rosettes consisting of six particles (cellulose-synthases) (**C**), which is never seen in the plasma membrane of non-growing cells (**A**) nor in those of mother semi-cells [14]. These cellulose-synthases are believed to be carried by special vesicles (flat vesicle) produced by Golgi bodies during the cellulose synthesis period. Hexagonal particle arrays and microfibril bands extended from these arrays are visible when cells are cultured in distilled water before freezing (**D** adapted from [15]). Cellulose-synthesizing rosettes were discovered in *Micrasterias* [16].

The figures show transmission electron micrographs of freeze fracture replicas of *Micrasterias crux-melitensis*, a zygote accumulating cellulose wall layers of *Closterium reinhardii*, and a fluorescence micrograph of growing *M. crux-melitensis* stained with calcofuor.

For freeze fracture, living cells mounted on supporting copper disks were frozen in liquid propane. Fracturing and shadowing were carried out at −112 °C. Cleansed replicas were imaged with an electron microscope at 100 kV. For fluorescence microscopy, cells were stained with 0.02 % calcofuor. Scale bars: 250 nm (**A**, **C**, **D**), 100 μm (**B**).

References

14. Noguchi T, Tanaka K, Ueda K (1981) Membrane structure of dictyosomes, large vesicles and plasma membranes in a green alga, *Micrasterias crux-melitensis*. Cell Struct Funct 6:217–229
15. Noguchi T, Ueda K (1985) Cell walls, Plasma membranes, and dictyosomes during zygote maturation of *Closterium ehrenbergii*. Protoplasma 128:64–71
16. Giddings Jr TH, Brower D, Staehelin LA (1980) Visualization of particle complexes in the plasma membrane of *Micrasterias denticulate* associated with the formation of cellulose fibrils in primary and secondary cell walls. J Cell Biol 84:327–339

Contributors

Tetsuko Noguchi*, Course of Biological Sciences, Faculty of Science, Nara Women's University, Kitauoya-nishimachi, Nara 630-8506, Japan
*E-mail: noguchi@cc.nara-wu.ac.jp

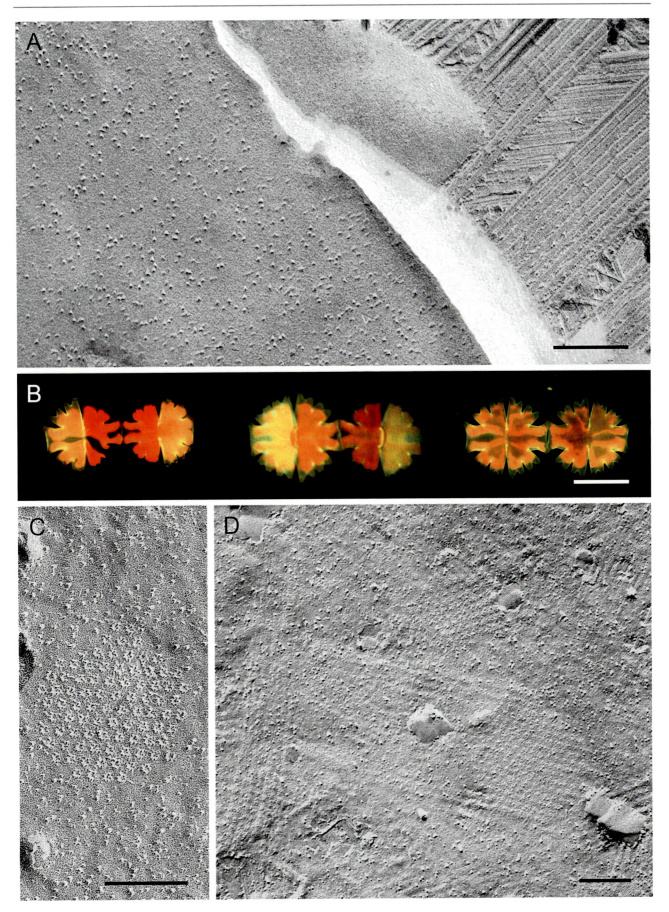

Plate 7.7

Localization of typical cell wall polysaccharides pectin and β-1,3/1,4 mixed linkage glucan in *Arabidopsis thaliana* and *Oryza sativa*

The structural and chemical nature of plant cell walls diversified extensively during the terrestrial invasion of plants in the Devonian period. The cell walls of land plants are characterized by two distinct layers of cell walls: the primary cell wall and the secondary cell wall, The former is ubiquitously found in all land plants, whereas the latter is found exclusively in vascular plants and is deposited onto the primary cell wall in specific tissue types, namely vascular tissues, after cell expansion has ceased.

The chemical nature of the primary cell wall has diversified among angiosperms. Although pectin and xyloglucans are abundant in the cell walls of eudicotyledonous plants, such as *Arabidopsis thaliana* (L.) Heynh [17], they are less abundant in the cell walls of Poales species, such as rice (*Oryza sativa* L.). Rice cell walls contain higher levels of β-1,3/1,4 mixed linkage glucans (MLG) and arabinoxylans than xyloglucans and pectin.

Pectin in the *A. thaliana* inflorescence stem (**A**) and MLG in *O. sativa* leaf blades (**B**) are shown by immunofluorescence. Green fluorescence indicates the presence of pectin or MLG. Similarly, pectin in cortical cells of the *A. thaliana* inflorescence stem (**C**) and MLG in *O. sativa* collenchyma cells of the leaf blade of (**D**) are shown by electron microscopy. Black (electron dense) particles indicate the presence of pectin or MLG within the cell wall.

For immunofluorescence analysis, stems or leaf blades were cross-sectioned using a vibratome at a thickness of 70 μm and fixed with 4 % paraformaldehyde in 20 mM sodium cacodylate buffer, pH 7.4, followed by probing with JIM5 monoclonal antibody for pectin or anti-MLG monoclonal antibody. JIM5 antibody specifically binds to a sparsely methylated or unesterified homogalacturonan domain of pectin, while the MLG antibody recognizes linear (1,3/1,4)-β-oligosaccharide segments in MLG. To enable highly sensitive immunodetection, the tyramide signal amplification method was used. For immunogold labeling, segments of stems or the leaf blades were fixed for 1 h in 0.1 M phosphate buffer, pH 7.2, containing 4 % paraformaldehyde and 0.05 % glutaraldehyde, then embedded in LR White resin (Sigma) and polymerized by heat. Fixed samples were subjected to ultra-thin sectioning followed by immuno-gold labeling with the appropriate monoclonal antibody for pectin or MLG.

Cor, cortex; Ph, phloem; If, interfascicular fiber; Ep, epidermal cell; Pa, parenchyma; Col, collenchyma. Scale bars: 100 μm (**A**), 30 μm (**B**), 5 μm (**C**), 1 μm (**D**).

Contributors

Ryusuke Yokoyama, Hideki Narukawa, Kazuhiko Nishitani*, Laboratory of Plant Cell Wall Biology, Graduate School of Life Sciences, Tohoku University, 6-3 Aoba, Aramaki, Aoba-Ku, Sendai, Miyagi 980-8578, Japan
*E-mail: nishitan@m.tohoku.ac.jp

References

17. Hongo S, Sato K, Yokoyama R, Nishitani K (2012) Demethylesterification of the primary wall by ECTIN METHYLESTERASE35 provides mechanical support to the Arabidopsis stem. Plant Cell 24:2624–2634

7 Cell Walls

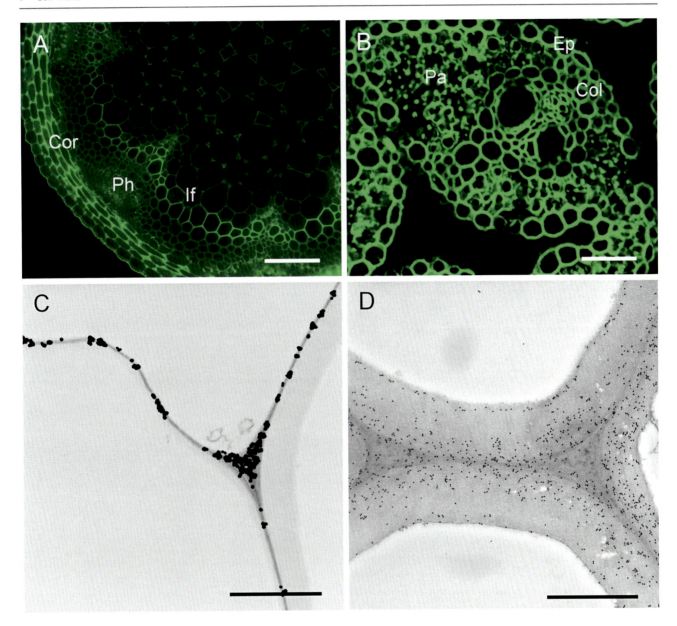

Plate 7.8

Plasmodesmata directly connect the cytoplasm of neighboring plant cells

Plant cells are surrounded by cell walls. Together, a pair of cell walls and the intervening lamella form an extracellular domain which separates neighbouring cells. Plasmodesmata penetrate both the primary and secondary cell walls, allowing transport of molecules between adjacent cells by symplast. Plasmodesmata also play important roles in cellular communication.

Membranes were stained by the osmium tetraoxide-potassium ferricyanide method (**A**). Plasmodesmata are tubular connections, 50–60 nm in diameter at the midpoint between adjacent cells [18]. The plasma membrane is continuous from one cell to the next cell at plasmodesmata. Plasmodesmata contain a desmotubule, which is a narrow tube-like structure (arrow heads) and is continuous with the smooth endoplasmic reticulum (ER) in the connected cells (**A**). A vertical cell wall section is imaged in samples prepared by high-pressure freezing and freeze substitution (**B**). Cortical microtubules can be clearly seen under the plasma membrane (arrows). The desmotubule does not fill the plasmodesma completely (**B**). The cytoplasmic sleeve exists between the plasma membrane and the desmotubule (arrow heads). Trafficking of molecules and ions through plasmodesmata occurs through this sleeve. Smaller molecules and ions can easily pass through plasmodesmata by diffusion without requiring additional chemical energy. Larger molecules, including proteins and RNA, can also pass through the cytoplasmic sleeve. It is known that in some cases molecule size restrictions can be overcome, though the mechanism of these transport systems is not yet understood. Special proteins and some viruses are able to increase the diameter of the channels enough for unusually large molecules to pass through [19]. A typical plant cell may have 10^3 to 10^5 plasmodesmata connecting to adjacent cells, which equates to 1 to 10 plasmodesmata per μm^2. There are two forms of plasmodesmata: primary plasmodesmata, which form during cell division when parental endoplasmic reticulum are trapped in the new cell wall [20], and secondary plasmodesmata, which form between mature cells. Members of the *Charophyceae*, *Charales*, *Coleochaetales* and *Phaeophyceae* algal families, as well as all embryphyte land plants, have plasmodesmata.

To stain ER and desmotubule, 5-day-old cotyledons were fixed in cacodylate buffer (Ph 7.4) containing 4 % paraformaldehyde, 1 % glutaraldehyde, and 0.1 M $CaCl_2$ for 3 h at 4 °C, washed with 0.1 M cacodylate buffer for 1.5 h, postfixed with 2 % OsO_4 plus 0.8 % $K_3Fe(CN)_6$ and 1 $\mu M CaCl_2$ in 0.1 M cacodylate buffer for 2 h at room temperature, and serially dehydrated in ethanol. To visualize desmotubules in the cell wall, cotyledons were frozen with a high-pressure freezer, then dehydrated for 2 days at −85 °C in acetone containing 3 % (w/v) osmium tetraoxide. Samples were embedded in Spurr resin, ultrathin sectioned, stained with uranium and lead, and imaged with an electron microscope. Scale bars: 1 μm.

Contributors

Yasuko Hayashi*, Department of Environmental Science, Graduate School of Science and Technology, Niigata University, Ikarashi, Niigata 950-2181, Japan
*E-mail: yhayashi@env.sc.niigata-u.ac.jp

References

18. Ding B, Turgeon R, Parthasarathy MV (1992) Substructure of freeze-substituted plasmodesmata. Protoplasma 169:28–41
19. Ding B (1998) Intercellular protein trafficking through plasmodesmata. Plant Mol Biol 38:279–310
20. Ehlers K, Kollmann R (2001) Primary and secondary plasmodesmata: structure, origin, and functioning. Protoplasma 216:1–30

7 Cell Walls 153

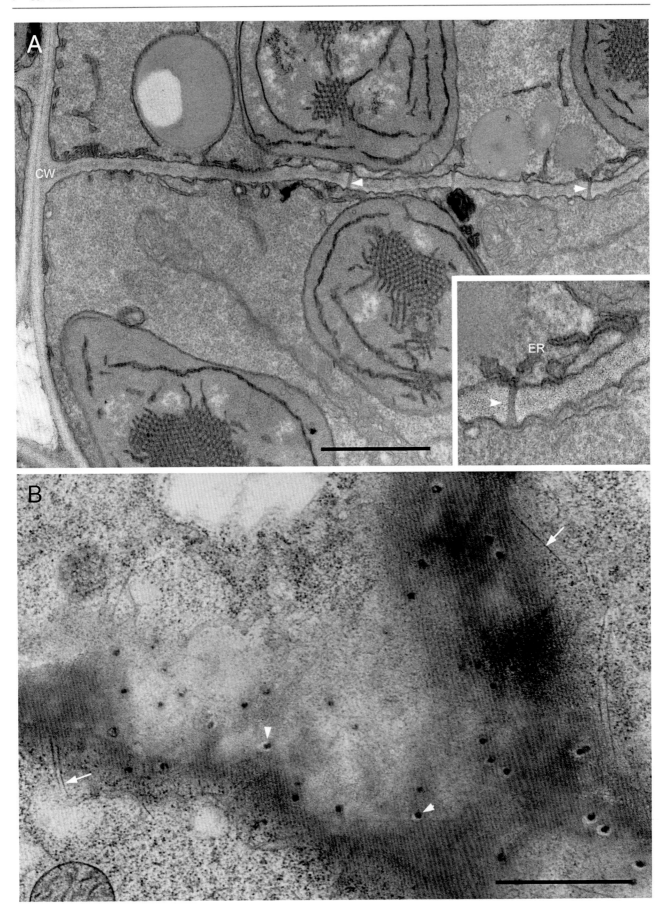

Plate 7.9

Meshwork structure of the Casparian strip

The Casparian strip is the barrier to apoplastic transport located in the radial walls of endodermal cells in roots and shoots of vascular plants. The Casparian strip is formed by individual endodermal cells, but must be continuous to function. The Casparian strip encircles each endodermal cell and is in fact continuous throughout the entire endodermal tissue. There is morphological evidence that, prior to lignification, positional information in the radial wall of endodermal cells defines the future site of strip formation.

This figure shows the meshwork structure of the Casparian strip. The main panel shows the hypocotyl of a 7 days old Arabidopsis seedling (*Arabidopsis thaliana* (L.) Heynh., Col-0) grown in the light (**A**). The Casparian strip and xylem vessels emit autofluorescence under UV light because they are impregnated with lignin (**A**). Three dimensional models of the Casparian strip in a hypocotyl display the tissue organization of a hypocotyl (**B**), the endodermis and xylem vessels (**C**), and the Casparian strip and xylem vessels (**D**). A Casparian strip isolated from a pea root (*Pisum sativum* L. cv. Alaska) observed under a fluorescence microscope is also shown (**E** adapted from [21]).

Isolation of the Casparian strip was performed as follows. Pea roots were split in half and incubated with a solution of cell wall digesting enzymes. Vascular tissues were removed from split roots and endodermal layers were picked up with forceps. Cells attached to isolated endodermal layers were removed and washed. Isolated endodermal layers were agitated on a Vortex mixer to remove any cytoplasm attached to the Casparian strips. Arabidopsis hypocotyls were fixed in FAA (3.7 % formaldehyde, 5 % acetic acid, and 50 % ethanol) and cleared in 10 % (w/v) KOH (105 °C, 1 min) (**A–D** adapted from [22]). Cleared hypocotyls were mounted on a glass slide and observed with a fluorescent microscope (BX-FLA; Olympus) equipped with a filter assembly for excitation by ultraviolet (UV) light (U-MWU: excitation filter, BP330-385; dichroic mirror, DM-400). Scale bar: 50 μm (**A**).

Contributors

Yoshihiro Honma, Ichirou Karahara*, Department of Biology, Graduate School of Science and Engineering, University of Toyama, 3190 Gofuku, Toyama 930-8555, Japan
*E-mail: karahara@sci.u-toyama.ac.jp

References

21. Karahara I (1995) The Casparian strip: the tight junction of plants. Kagaku to Seibutsu 33:246–251 (Japanese)
22. Homma Y, Karahara I (2004) The Casparian strip in a cleared hypocotyl of Arabidopsis. Cytologia 69:i–ii (Technical note)

7 Cell Walls 155

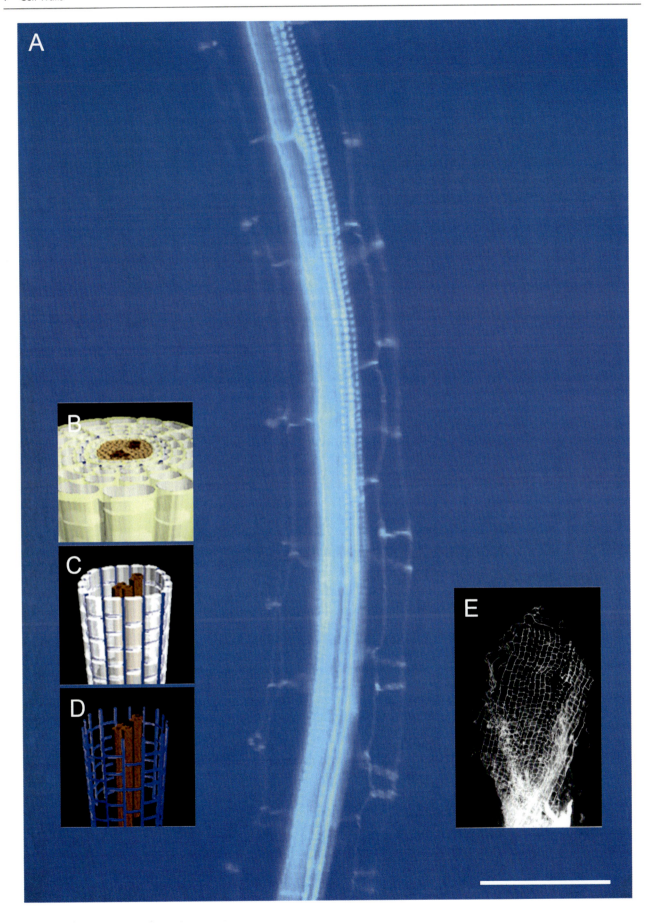

Chapter References

1. Osumi M (1998) The ultrastructure of yeast: cell wall structure and formation. Micron 29:207–233
2. Osumi M, Yamada N, Kobori H, Naito N, Baba M, Nagatani T (1989) Cell wall formation in regenerating protoplasts of *Schlzosaccharomyces pombe*. J Electron Microsc 38:457–468
3. Osumi M (2012) Visualization of yeast cells by electron microscopy. J Electron Microsc 61:343–365
4. Osumi M, Yamada N, Yaguchi H, Kobori H, Nagatani T, Sato M (1995) Ultrahigh-resolution low-voltage SEM reveals ultrastructure of the glucan network formation fission yeast protoplast. J Electron Microsc 44:198–206
5. Yamamoto M, Fujishita M, Hirata A, Kawano S (2004) Regeneration and maturation of daughter cell walls in the autospore-forming green alga *Chlorella vulgaris* (Chlorophyta, Trebouxiophyceae). J Plant Res 117:257–264
6. Yamamoto M, Kawano S (2004) Daughter cell wall synthesis in autorpore-forming green alga, *Chlorella vulgaris* (IAM-C-536). Cytologia 69–3:i–ii
7. Yamamoto M, Nishikawa T, Kajitani H, Kawano S (2007) Patterns of asexual reproduction in *Nannochloris bacillaris* and *Marvania geminata* (Chlorophyta, Trebouxiophyceae). Planta 226:917–927
8. Yamazaki T, Owari S, Ota H, Sumiya N, Yamamoto M, Watanabe K, Nagumo T, Miyamura S, Kawano S (2013) Localoization and evolution of septins in algae. Plant J 74:605–614
9. Sekida S, Horiguchi T, Okuda K (2001) Development of the cell covering in the dinoflagellate *Scrippsiella hexapraecingula* (Peridiniales, Dinophyceae). Phycological Res 49:163–176
10. Sekida S, Horiguchi T, Okuda K (2004) Development of thecal plates and pellicle in the dinoflagellate *Scrippsiella hexapraecingula* (Peridiniales, Dinophyceae) elucidated by changes in stainability of the associated membranes. Eur J Phycol 39:105–114
11. Sekida S, Takahira M, Horiguchi T, Okuda K (2012) Effects of high pressure in the armored dinoflagellate *Scrippsiella hexapraecingula* (Peridiniales, Dinophyceae): changes in thecal plate pattern and microtubule assembly. J Phycol 48:163–173
12. Ueda K, Noguchi T (1976) Transformation of the Golgi apparatus in the cell cycle of a green alga, *Micrasterias americana*. Protoplasma 87:145–162
13. Ueda K, Yoshioka S (1976) Cell wall development of *Micrasterias americana*, especially in isotonic and hypertonic solutions. J Cell Sci 21:617–631
14. Noguchi T, Tanaka K, Ueda K (1981) Membrane structure of dictyosomes, large vesicles and plasma membranes in a green alga, *Micrasterias crux-melitensis*. Cell Struct Funct 6:217–229
15. Noguchi T, Ueda K (1985) Cell walls, plasma membranes, and dictyosomes during zygote maturation of *Closterium ehrenbergii*. Protoplasma 128:64–71
16. Giddings Jr TH, Brower D, Staehelin LA (1980) Visualization of particle complexes in the plasma membrane of *Micrasterias denticulate* associated with the formation of cellulose fibrils in primary and secondary cell walls. J Cell Biol 84:327–339
17. Hongo S, Sato K, Yokoyama R, Nishitani K (2012) Demethylesterification of the primary wall by PECTIN METHYLESTERASE35 provides mechanical support to the Arabidopsis stem. Plant Cell 24:2624–2634
18. Ding B, Turgeon R, Parthasarathy MV (1992) Substructure of freeze-substituted plasmodesmata. Protoplasma 169:28–41
19. Ding B (1998) Intercellular protein trafficking through plasmodesmata. Plant Mol Biol 38:279–310
20. Ehlers K, Kollmann R (2001) Primary and secondary plasmodesmata: structure, origin, and functioning. Protoplasma 216:1–30
21. Karahara I (1995) The Casparian strip: the tight junction of plants. Kagaku to Seibutsu 33:246–251 (Japanese)
22. Homma Y, Karahara I (2004) The Casparian strip in a cleared hypocotyl of Arabidopsis. Cytologia 69:i–ii (Technical note)

Generative Cells

8

Atsushi Sakai

Sexual reproduction is one of the most important events in the life history of eukaryotic organisms. Throughout the process of algal and plant evolution, the mode of mating has evolved from isogamy to unisogamy, and, finally, oogamy. While all three types are seen among algae, only oogamy is found among land plants.

Across this evolutionary process, mating becomes progressively less dependent on the presence of external water. In algae living in water, gametes can move easily toward other gametes by swimming in external water. In basal land plants, such as bryophytes and pteridophytes, sperms also move from antheridia to eggs within the archegonia by swimming in external water. In some gymnosperms, male gametes are still sperms with flagella, but they do not reach eggs by swimming through external water; rather, pollen grains (male gametophytes encapsulated within a pollen wall) travel through dry air to ovules, and develop therein to form and release sperms. Thus, the sperms must swim only a short distance across the fluid filled in archegonial chamber within the ovule. Later, during the evolution of gymnosperms, male gametes lost flagella and motility, and are delivered by pollen tubes, which make it possible for the male gametes to reach egg cells without being exposed to the extracellular environment. In angiosperms, pollen grains land on the stigma and extend pollen tubes through the style to deliver sperm cells to ovules, making the process almost independent of external fluid.

Another trend in plant evolution is the simplification of gametophyte generation. In bryophyte, gametophyte generation is the dominant phase in life cycle. Small body sizes are preferred for terrestrial gametophytes because antheridia and archegonia must be connected by a water corridor for fertilization to occur. Pteridophytes, therefore, keep gametophytes (prothallia) small while sporophytes evolve to be larger and more complex. In gymnosperms, sporophytes become more dominant, while gametophytes become smaller; the male gametophyte (pollen and its derivatives) is composed of relatively few cells while the female gametophyte (embryo sac) is composed of several thousand cells and is embedded in (and dependent on) sporophyte tissues. In angiosperms, the male gametophyte (pollen) is composed of only three cells (one vegetative cell and two sperm cells), and the female gametophyte (embryo sac) is typically composed of seven cells (one central cell, one egg cell, two synergids and three antipodal cells). Among the ten gametophytic cells, however, as many as four cells (two sperm cells, one egg cell and one central cell) participate in the double fertilization specific to angiosperms.

The 14 selected figures in this chapter illustrate various aspects of sexual reproduction in algae and plants. Y. Mogi and S. Kawano illustrate asymmetrical mating in the sea weed *Ulva compressa*, which represents the transition from isogamy to unisogamy. K. Ueda and T. Noguchi show the process of flagellated sperm formation in a liverwort. H. Nishida and S. Miyamura demonstrate multi-flagellated, motile sperms in an extinct fossil plant *Glossopteris* and an extant "living fossil" ginkgo, respectively. N. Mogami and I. Tanaka show structures of a pollen wall and an encapsulated male gametophyte in lily. C. Saito and K. Shoda demonstrate differences in the ultrastructures between two kinds of gametophytic cells in young *Arabidopsis* pollen. N. Nagata reports that selective and independent replication or degradation of plastid and mitochondrial DNAs in young generative cells is important for determining the pattern of cytoplasmic inheritance in bananas. C. Saito presents dimorphic sperm cells in *Plumbago auriculata*, which suggests the division of roles between the two sperm cells. M. M. Kanaoka illustrates pollen tube guidance toward the ovules in vivo in *Arabidopsis* pistil. T. Higashiyama introduces the pollen tube guidance and fertilization process in angiosperms using a semi-in vitro *Trenia* system. M. M. Kanaoka illustrates characteristics of the *Torenia* (and *Lindernia*) ovules with partially protruded embryo sacs, and demonstrates the isolation of protoplasts from gametophytic cells. H. Kuroiwa and T. Kuroiwa report changes in the size, shape, and DNA content of mitochondria in egg cells of geranium, as well as dynamic changes in the ultrastructure of zygotes, synergids, and sperm cells at very early stages of embryogenesis. Finally, Y. Mineyuki et al. visualize cell geometry in a whole *Arabidopsis* seed, the final product of sexual reproduction in seed plants, using novel X-ray micro-CT technology.

A. Sakai (✉)
Course of Biological Sciences, Faculty of Science, Nara Women's University, Kitauoya-nishimachi, Nara 630-8506, Japan
e-mail: sakai@cc.nara-wu.ac.jp

T. Noguchi et al. (eds.), *Atlas of Plant Cell Structure*,
DOI 10.1007/978-4-431-54941-3_8, © Springer Japan 2014

Plate 8.1

A mating-pair of seaweed *Ulva compressa* gametes with asymmetrical mating structure positions

The green macroalgal genus *Ulva*, which is widely distributed in marine and fresh waters all over the world, includes several economically valuable species. In laboratory culture, *U. compressa* completes its life cycle within 6 weeks and gametogenesis is easily induced, which enables synchronized harvesting of gametes at specific developmental stages. *U. compressa* gametes are biflagellate and pear shaped, with one eyespot at the posterior end of the cell [1]. The species is at an early evolutionary stage of sexual reproduction, between isogamy and anisogamy. In the past, zygote formation of green algae was categorized solely by the relative sizes of gametes produced by the two mating types (+ and −). Recently, however, asymmetry of gamete cell fusion sites and mating structure positions have been observed to differ between mating types in several green algae.

A new method visualizing mating structures of *U. compressa* by field emission scanning electron microscopy (FE-SEM) was used to apply this asymmetry for determining gamete mating types [2]. When gametes were subjected to drying stress in the process of a conventional critical-point-drying method, a round structure was observed on cell surfaces. In mt+, this structure was located on the same side of the cell as the eyespot, whereas it was on the side opposite the eyespot in mt−. Gametes fuse at the round structures (arrow in the inset). Transmission electron microscopy (TEM) showed alignment of vesicles inside the cytoplasm directly below the round structures, which are indeed the mating structures. Serial sectioning and three-dimensional reconstruction of TEM micrographs confirmed the association of the mating structure with flagella roots. The mating structure was associated with 1 d root in the mt+ gamete but with 2 d root in the mt− gamete. This asymmetry of cell fusion sites and/or mating structure positions is relevant to the sexuality of the two gametes.

Field emission scanning electron micrographs show zygotes in the early phase of cell fusion and an enlarged image of the cell fusion site. E, eyespot. Scale bar: 1 µm.

Contributors

Yuko Mogi, Shigeyuki Kawano*, Department of Integrated Biosciences, Graduate School of Frontier Sciences, The University of Tokyo, Bldg. FSB-601, 5-1-5 Kashiwanoha, Kashiwa, Chiba 277-8562, Japan
*E-mail: kawano@k.u-tokyo.ac.jp

References

1. Kagami Y, Mogi Y, Arai T, Yamamoto M, Kuwano K, Kawano S (2008) Sexuality and uniparental inheritance of chloroplast DNA in the isogamous green alga *Ulva compressa* (Ulvophyceae). J Phycol 44:691–702. doi:10.1111/j.1529-8817.2008.00527.x
2. Mogi Y, Kagami Y, Kuwano K, Miyamura S, Nagumo T, Kawano S (2008) Asymmetry of eyespot and mating structure positions in *Ulva compressa* (Ulvales, Chlorophyta) revealed by a new FE-SEM method. J Phycol 44:1290–1299. doi:10.1111/j.1529-8817.2008.00573.x

8 Generative Cells

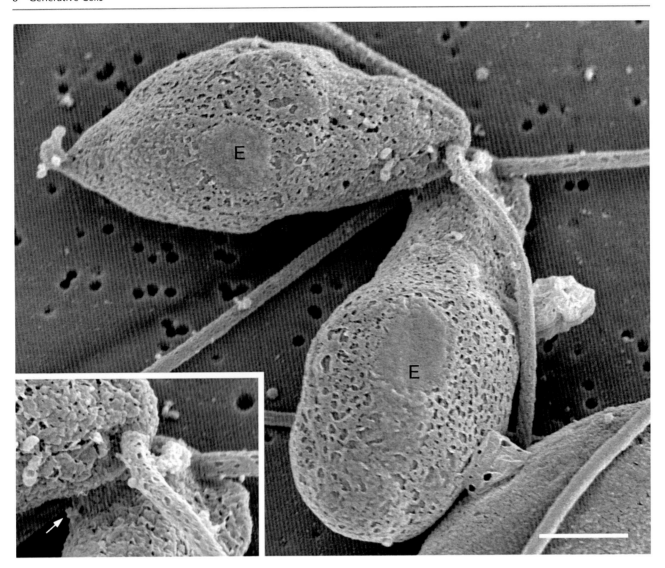

Plate 8.2

Spermatogenesis in *Marchantia polymorpha*

The sex organs of *Marchantia*, antheridia (male) and archegonia (female), are made on the male and female receptacles, respectively, and formed on the thalli of gametophores (n). A subpopulation of antheridium cells differentiate into spermatogenic cells, which subsequently divide mitotically and form a great number of spermatids. These cells dynamically change shape and become sperms, each of which has two flagella. In the presence of external water, sperms then swim to the egg cells in the archegonia and fertilize them (oogamy). Fertilized eggs (zygotes, 2n) develop into sporophytes.

Spermatogenic cells have a large nucleus (**A**). After they become spermatids, their cytoplasm decreases in volume due to digestion of organelles by lysosomes (L in **B**) while cell walls maintain their original shapes. As a result, the space between the cell wall and the plasma membrane increases in volume, and this is where digested materials from the cytoplasm accumulate. During this process, nuclei elongate concomitant with the disappearance of nucleoli. Two flagella protrude from the plasma membrane and surround the cell (**B**). As cytoplasm is digested, nuclei elongate further while increasing in electron density (**C**). Finally, the cells become sperm cells, each with a large nucleus and two flagella. At this time, cell walls are digested and the sperm cells are easily released from the antheridium.

Figures (**A–C**) show transmission electron micrographs of antheridium cells of *Marchantia polymorpha* L. An artificial image of a flagellum cross section is also shown (**inset** in **B**). Antheridia were fixed with 3 % glutaraldehyde and post-fixed with 1 % OsO_4. After postfixation cells were dehydrated using a graded acetone series, and infiltrated with increasing concentrations of Spurr resin. Ultrathin sections were stained with lead citrate and images were obtained using a transmission electron microscope at 80 kV. CW, cell wall; N, nucleus; L, lysosome; arrowhead, flagellum; Scale bars: 1 μm (**main**), 0.1 μm (**inset** in **B**).

Contributors

Katsumi Ueda[1], Tetsuko Noguchi[2]*, [1]Deceased, [2]Course of Biological Sciences, Faculty of Science, Nara Women's University, Kitauoya-nishimachi, Nara 630-8506, Japan
*E-mail: noguchi@cc.nara-wu.ac.jp

8 Generative Cells

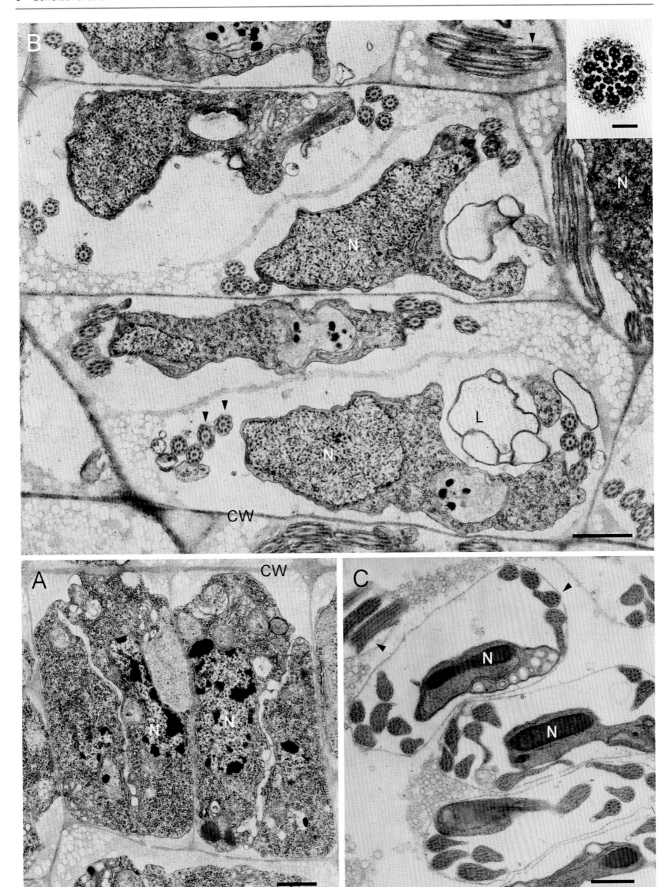

Plate 8.3

Motile sperms released in the ovule of an extinct Permian gymnosperm *Glossopteris*

A fossil of a motile sperm released in the ovule of an extinct Gondwanan plant, *Glossopteris*, was discovered in megasporangiate organs preserved in silicified peat of late Permian age (ca. 250 million years ago) from the Bowen Basin in Queensland, Australia [3, 4]. *Glossopteris* is well known as a dominant plant group endemic to the super-continent Gondwana. The plant is believed to have been a deciduous tree that covered the marshland formed at the periphery the continental ice sheet which overlaid mainland Gondwana. *Glossopteris* fossils are usually found as impressions of leaves and detached monosporangiate reproductive organs which are classified in different morphotypes customarily described as separate morphogenera, e.g., *Denkania*, *Dictyopteridium* and *Scutum* for megasporangiate organs, and *Eretmonia* and *Glossotheca* for microsporangiate organs. In spite of its reputation, *Glossopteris* is not well characterized morphologically because of the paucity of fossils suitable for anatomical study. The anatomical structures of reproductive organs and the reproductive biology of *Glossopteris*, therefore, are poorly understood.

The Homevale locality of Queensland is famous for outcropping permineralized fossils of well-preserved *Glossopteris* remains. Motile sperms were found in a *Dictyopteridium*-type megasporangiate organ which was later described as a new morphogenus of permineralized organ *Homevaleia* [5]. Serial cross sections of *Homevaleia gouldii* show numerous ovules born on the adaxial surface of a megasporangiate leaf (**A, B**). Arrowheads indicate the ovule, where motile sperms are found. The ovule has a megagametophyte producing a single archegonium containing one egg cell (**A**) and several pollen tubes in the pollination chamber (**B**). A composite image compiled from serial sections of the ovule (**C**) shows both pollen tubes and the egg cell together, though the megagametophyte and egg cell images are taken from a section not shown (**A** or **B**). Each pollen tube is identified by the number in the figure. Two dark granules in pollen tube #1 may be unreleased sperms. Two arrows indicate sperms released from pollen tube #2. One released sperm is about 10 µm in diameter, much smaller than sperm of *Ginkgo* and Cycadophytes (**D**). White dots may represent basal bodies of flagella born on a helical blepharoblast. Estimated number of flagella per sperm is approximately 55 (**E**), which is close to the number of flagella found on fern sperms and far smaller than the number on *Ginkgo* (ca. 1,000) or cycads (20,000 to >60,000). This is the third evidence of zooidogamy in seed plants, in addition to previous discoveries in living *Ginkgo* and Cycadophytes, both of which are famous achievements of early Japanese plant morphology studies in the late nineteenth century. Scale bars: 1 mm (**B**), 0.1 mm (**C**), 5 µm (**D**).

Contributors

Harufumi Nishida*, Department of Biological Sciences, Faculty of Science and Engineering, Chuo University, 1-13-27 Kasuga, Bunkyo, Tokyo 112-8551, Japan
*E-mail: helecho@bio.chuo-u.ac.jp

References

3. Nishida H, Pigg KB, Rigby JF (2003) Swimming sperm in an extinct Gondwanan plant. Nature 422:396–397. doi:10.1038/422396a
4. Nishida H, Pigg KB, Kudo K, Rigby JF (2004) Zooidogamy in the late Permian genus *Glossopteris*. J Plant Res 117:323–328. doi:10.1007/s10265-004-0164-4
5. Nishida H, Pigg KB, Kudo K, Rigby JF (2007) New evidence of reproductive organs of *Glossopteris* based on permineralized fossils from Queensland, Australia. I. Ovulate organ *Homevaleia* gen nov. J Plant Res 120:539–549. doi:10.1007/s10265-007-0093-0

8 Generative Cells 163

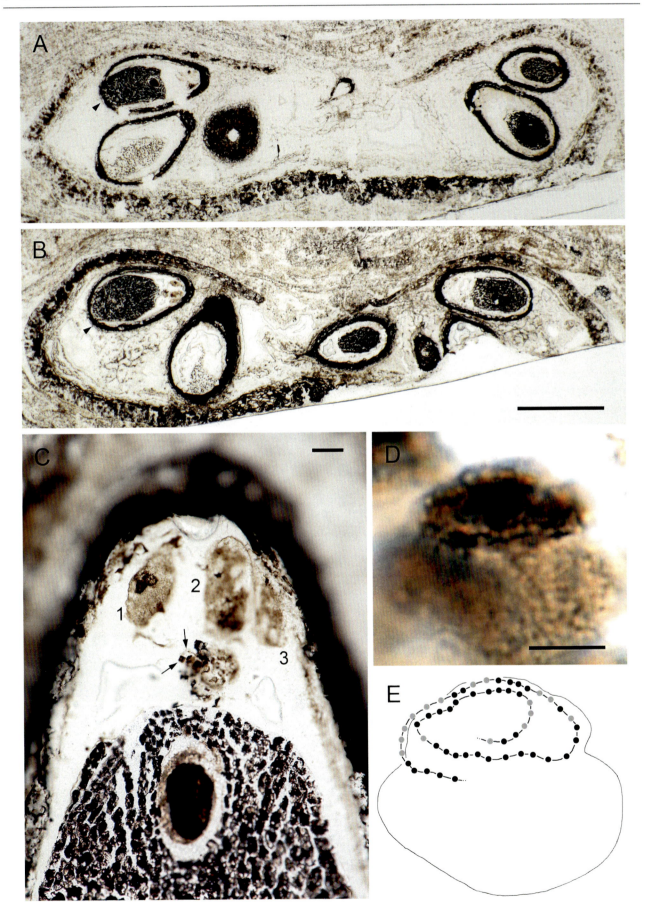

Plate 8.4

Multiflagellated sperm of *Ginkgo biloba* L

The sperms of *Ginkgo biloba* L. and cycads are the only flagellate male gametes in extant seed plants [6]. In 1896 the Japanese botanist Sakugoro Hirase first observed living *Ginkgo* sperms using a light microscope and described them as ellipsoid in form with a spiral band of numerous flagella at their anterior end. *Gingko* sperms are formed once a year between the end of August and middle of September. During this season, sperms are found in the pollen tube (male gametophyte) anchored in the sporophyte tissue of the ovule attached to the female tree. The sperms are released from the anterior end of the pollen tube into the archegonial chamber filled with fluid and swim to the egg cell in the archegonium using flagella. After passing through a small gap between neck cells at the top of the archegonium, one of the sperms fuses with the egg cell. Each sperm has thousands of flagella arranged in a spiral band. The flagellar beat is initiated in the center of the spiral, and travels along the spiral from the center to the outside. Flagella extend from basal bodies at the base of the flagellar spiral band. The spiral form of the anterior part of the sperm is shaped by an array of spline microtubules which originates from the multilayered structure (MLS), which is a characteristic of streptophyte flagellate cells.

Ginkgo sperm are shown in a scanning electron microscope image in this figure. Sperms released from the pollen tube were picked up with a micropipette, fixed with 3 % glutaraldehyde, and post-fixed with 1 % OsO_4. After dehydration through a graded ethanol series, cells were critical point dried, coated with platinum-palladium and observed with scanning electron microscope at 15 kV. Scale bar: 10 μm.

Contributors

Shinichi Miyamura*, Faculty of Life and Environmental Sciences, University of Tsukuba, Tsukuba, Ibaraki 305-8572, Japan
*E-mail: miyamura.shinichi.fw@u.tsukuba.ac.jp

References

6. Hori T, Miyamura S (1997) Contribution to the knowledge of fertilization of gymnosperms with flagellated sperm cells: *Ginkgo biloba* and *Cycas revoluta*. In: Hori T, Ridge RW, Tulecke W, Tredici DP, Tremouillaux-Guiller J, Tobe H (eds) *Ginkgo biloba* a global treasure: From biology to medicine. Springer-Verlag, Tokyo, pp 67–84. doi:10.1007/978-4-431-68416-9_6

8 Generative Cells

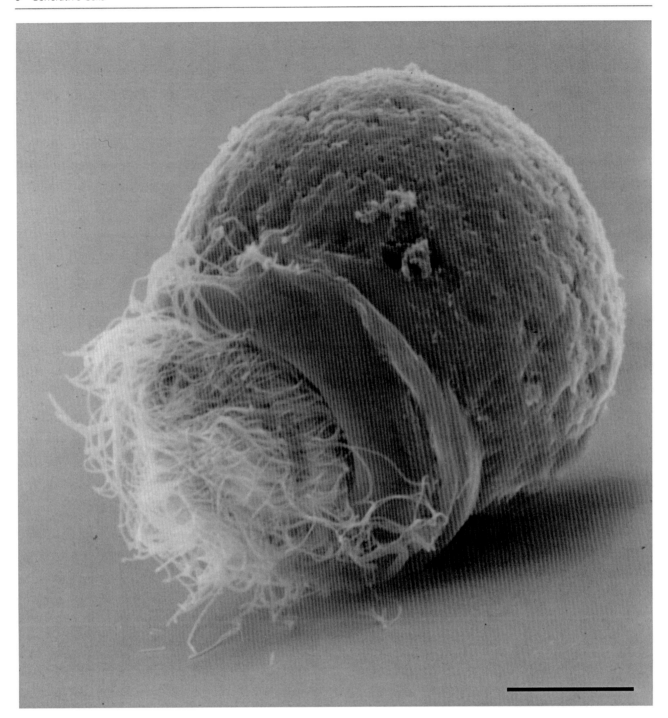

Plate 8.5

Pollen exine and male gametic nucleus of *Lilium longiflorum*

Pollen have a specific structure, which includes the exine, a resistant outer wall. Exine formation begins just prior to separation of the tetrad after meiosis, but does not take place in future aperture regions. During exine wall development, sporopollenin, the chemical material characteristic of the exine, is deposited. As a result, the four microspores released from the tetrad can be characterized by a species-specific sculptured exine and aperture, as shown in four microspores of *Lilium longiflorum* which are stained with 1 % auramine O and observed under an epifluorescence microscope (**A**) [7]. During microspore and pollen development, additional lipoidal and pigmented substances accumulate on the exine. Protoplasts can be directly released from the maturing lily pollen through a break in the aperture devoid of sculptured exine.

Pollen, the male gametophyte of angiosperms, consists of two functionally and structurally different types, either one generative cell plus one vegetative cell (bicellular type), or two sperm cells within one vegetative cell (tricellular type). The differentiation of these cells is accompanied by an asymmetric cell division (microspore mitosis) which partitions the mother cell into two daughters of unequal size. Thereafter, the smaller, generative cell comes to be embedded in the cytoplasm of the larger, vegetative cell. The generative cell is the male gametic cell, which produces two sperm cells either before or after germination for subsequent double fertilization. Meanwhile, the non-reproductive vegetative cell is believed to function in pollen germination and pollen-tube growth. Although the pollen-specific arrangement of cells (a generative cell within a vegetative cell) can be observed by electron microscopy, it is generally difficult to demonstrate the presence of the male gametic cell(s) in pollen by ordinary light microscopy. However, in protoplasts isolated from pollen of *Lilium longiflorum*, the cell-within-a-cell arrangement is clearly visible by propionic orcein (1 %) staining (**B**) [8]. The nucleus of the spindle-shaped generative cell has highly condensed chromatin, as do the nuclei of the two sperm cells, whereas the nucleus of the vegetative cell contains diffuse chromatin. Detailed biochemical studies comparing between those morphologically different nuclei (heterochromatic nuclei of generative cells and euchromatic nuclei of vegetative cells) have characterized novel histone variants (gH2A, gH2B and gH3), which are specific to male gametic nuclei and may be related to condensation of chromatin in the male gametic nucleus of *Lilium longiflorum* [9]. Asterisks, apertures; GC, generative cell; N^{-GC}, nucleus of generative cell; N, nucleus of vegetative cell. Scale bars: 50 μm (**A**), 25 μm (**B**).

Contributors

Norifumi Mogami[1], Ichiro Tanaka[2]*, [1]Department of Biological and Chemical Systems Engineering, Kumamoto National College of Technology, Kumamoto 866-8501, Japan, [2]Graduate School of Nanobioscience, Yokohama City University, Seto 22-2, Kanazawa-ku, Yokohama, Kanagawa 236-0027, Japan
*E-mail: itanaka@yokohama-cu.ac.jp

References

7. Tanaka I (2011) Molecular morphological studies on pollen development using protoplasts. Plant Morphol 23:53–59
8. Tanaka I (1993) Development of male gametes in flowering plants. J Plant Res 106:55–63. doi:10.1007/BF02344373
9. Ueda K, Tanaka I (1995) Male gametic nucleus-specific H2B and H3 histones, designated gH2B and gH3, in *Lilium longiflorum*. Planta 197:289–295. doi:10.1007/BF00202649

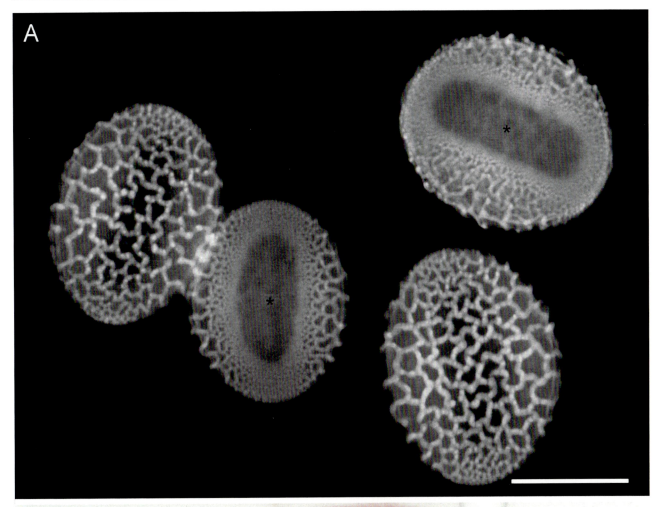

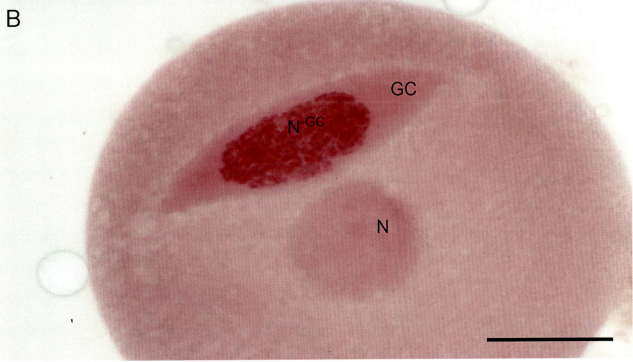

Plate 8.6

Developing *Arabidopsis* pollen grain containing a young generative cell with some mitochondria and no plastids

Pollen mitosis I, which occurs just after meiosis, is a significant asymmetric division which produces a large vegetative cell and a small generative cell. In *Arabidopsis thaliana*, young generative cells contain few mitochondria and no plastids just after pollen mitosis I. The generative cell further divides to give rise to two sperm cells before fertilization. Eventually, almost no mitochondria nor plastids can be detected in *Arabidopsis* sperm cells. *A. thaliana*, like approximately 80 % of angiosperm species, shows maternal inheritance of plastids and mitochondria. The exclusion of plastids and mitochondria from generative and sperm cells might be responsible, at least in part, for maternal inheritance of organelles in *Arabidopsis*.

A section of developing Arabidopsis pollen grain shows a small, spindle-shaped generative cell (GC) embedded in a large vegetative cell. While the vegetative cell contains a large, irregular-shaped nucleus (N) with nucleolus and dispersed chromatin, mitochondria (M), and amyloplasts (Am) containing large starch grains, the generative cell contains a small, condensed nucleus (N^{-GC}), few mitochondria, and no plastids.

Developing anthers were collected days before flowering and rapidly frozen in a high-pressure freezer. Frozen samples were then transferred to 2 % OsO_4 in anhydrous acetone and kept at $-80\ °C$ for 4 days. Samples were held at $-20\ °C$ for 2 h, 4 °C for 2 h, and then at room temperature for 10 m. After several washes with anhydrous acetone, the samples were embedded in Spurr resin, cut into ultrathin sections, stained, and imaged under a transmission electron microscope. Scale bar: 1 μm.

Contributors

Chieko Saito[1]*, Keiko Shoda[2], [1]Department of Biological Sciences, Graduate School of Science, The University of Tokyo, 7-3-1 Hongo, Bunkyo-ku, Tokyo 113-0033, Japan, [2]Laboratory for Cell Function Dynamics, RIKEN Brain Science Institute, 2–1 Hirosawa, Wako, Saitama 351-0198, Japan
*E-mail: chiezo@bs.s.u-tokyo.ac.jp

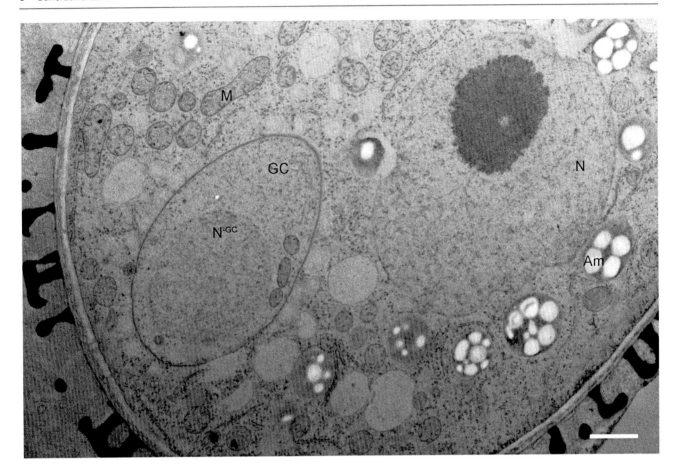

Plate 8.7

The selective increase or decrease of organelle DNAs in young generative cells controls cytoplasmic inheritance in higher plants

Since Correns and Baur first described the non-Mendelian inheritance of leaf color there have been many reports of cytoplasmic inheritance of organelles in higher plants. Cytological observations of mature pollen grains from over 500 angiosperm species by 4′,6-diamidino-2-phenylindole (DAPI)-fluorescence microscopy revealed a close correlation between the cytologically determined presence/absence of organellar DNA in generative/sperm cells and genetically determined biparental/maternal inheritance of plastids. Cytological observations, however, did not distinguish mitochondrial DNA (mtDNA) from plastid DNA (ptDNA). To distinguish plastid DNA and mitochondrial DNA in generative/sperm cells, the DAPI-DiOC$_6$ double-staining of Technovit sections [10] was applied to pollen grains from eight angiosperm species whose modes of plastid and mitochondrial inheritance had been determined genetically [11]. The eight species were classified into four types (p$^+$m$^+$, p$^+$m$^-$, p$^-$m$^+$, and p$^-$m$^-$) based on the presence or absence of ptDNA and mtDNA in mature generative cells, and this cytological classification was in complete agreement with genetically determined patterns of organelle inheritance. Moreover, the presence or absence of organelle DNA was determined at least in part by a drastic increase or decrease in the DNA content of the respective organelle just after microspore mitosis I. Thus, the mode of cytoplasmic inheritance in higher plants appears to be determined in young generative cells by independently controlling the behavior of mtDNA and ptDNA.

Fluorescence images of mature generative cells of bananas, *Musa acuminate*, a (p$^-$m$^+$) type plant, stained with DiOC$_6$ (**A**) and DAPI (**B**) are shown in this figure. Each mitochondrion (a bright fluorescent spot in DiOC$_6$ stained cells) has a single intense spot of DAPI fluorescence, while each plastid (an array of black areas representing starch grains) has no DAPI fluorescence, indicating that *M. acuminata* generative cells include mtDNA but no ptDNA. This is in line with genetic analysis, which suggests paternal inheritance of mtDNA and maternal inheritance of ptDNA in *M. acuminate* [12].

Anthers were fixed with 2 % glutaraldehyde buffered with 20 mM sodium cacodylate at pH 7.2 for 16 h at 4 °C, dehydrated through an alcohol series, and embedded in Technovit 7100 resin. Thin sections (0.6 μm thick) were cut with a glass knife on an ultramicrotome, dried on coverslips, stained with 100 μg/mL DiOC$_6$ in ethanol for 10 s, and washed with distilled water for 10 s. Sections were further stained with an equal volume of 1 μg/mL DAPI in TAN buffer. Samples were imaged with an epifluorescence microscope. Scale bar: 5 μm.

Contributors

Noriko Nagata*, Faculty of Science, Japan Women's University, 2-8-1 Mejirodai, Bunkyo-ku, Tokyo 112-8681, Japan
*E-mail: n-nagata@fc.jwu.ac.jp

References

10. Fujie M, Kuroiwa H, Kawano S, Mutoh S, Kuroiwa T (1994) Behavior of organelles and their nucleoids in the shoot apical meristem during leaf development in *Arabidopsis thaliana* L. Planta 194:395–405. doi:10.1007/BF00197541
11. Nagata N, Saito C, Sakai A, Kuroiwa H, Kuroiwa T (1999) The selective increase or decrease of organellar DNA in generative cells just after pollen mitosis one controls cytoplasmic inheritance. Planta 209:53–65. doi:10.1007/s004250050606
12. Fauré S, Noyer JL, Carreel F, Horry JP, Bakry F, Lanaud C (1994) Maternal inheritance of chloroplast genome and paternal inheritance of mitochondrial genome in bananas (*Musa acuminata*). Curr Genet 25:265–269. doi:10.1007/BF00357172

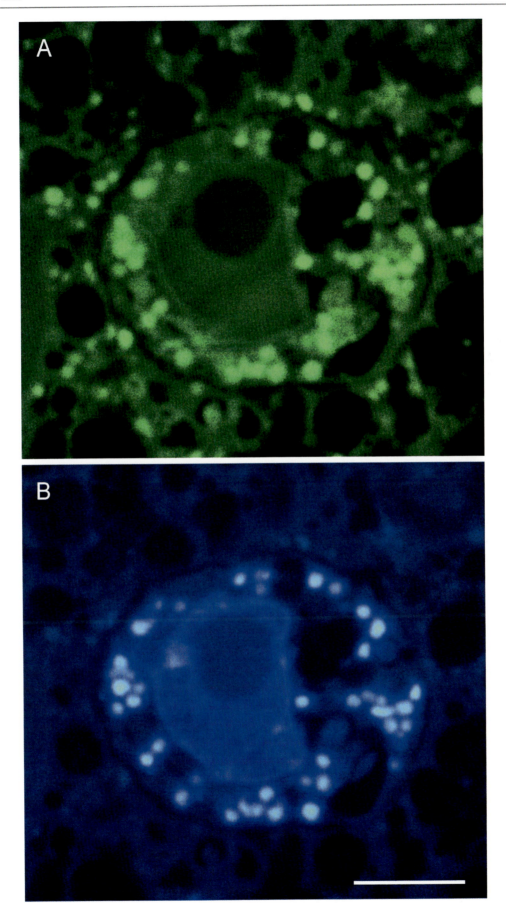

Plate 8.8

Dimorphic *Plumbago auriculata* sperm cells

Some angiosperm species, such as *Plumbago zylanica*, produce a clearly distinguishable dimorphic pair of sperm cells [13]. In *P. zeylanica*, the sperm cell associated with the vegetative nucleus (S_{vn}) is mitochondria-rich, and the sperm cell unassociated with the vegetative nucleus (S_{ua}) is plastid-rich. To date, *P. zeylanica* is the only clear example of the preferential fertilization; while S_{ua} preferentially fuses with the egg cell, S_{vn} fuses with the central cell [14].

As mitochondria and plastids in *Plumbago* sperm cells contain DNA, they can be visualized and distinguished from each other by 4′, 6-diamidino-2-phenylindole (DAPI)-fluorescent microscopy [15]. This technique is easier than traditional electron microscopy. A pair of sperm cells in *P. auriculata*, a commercially available species closely related to *P. zeylanica*, imaged by DAPI-fluorescent microscopy the dimorphism of *Plumbago* sperm cells (**A**). While the right sperm cell (Sua) contains a number of relatively large fluorescent spots (plastid nucleoids) in the cytoplasm, the left sperm cell (Svn) contains only extremely small fluorescent spots (mitochondrial nucleoids). The positions of the dimorphic sperm cells relative to the vegetative nucleus are shown in a 3D-reconstructed image of a pair of sperm cells from a mature pollen grain of *P. auriculata* (**B**).

For DAPI analysis, anthers were excised from *P. auriculata* flowers, fixed in 2 % glutaraldehyde in TAN buffer (20 mM Tris–HCl, pH 7.6. 0.5 mM EDTA, 7 mM 2-mercaptoethanol, 1.2 mM spermidine), and dissected in the fixing solution with a blade and tweezers. A suspension of fixed samples was then stained with 1 μg/mL DAPI, mounted with 1 mg/mL of n-propyl gallate dissolved in 50 % glycerol, squashed adequately, and observed with an epi-fluorescence microscope. For 3D reconstruction, anthers were fixed in 2 % glutaraldehyde, dehydrated by a series of ethanol washes, and embedded in Technovit 7100 resin. Serial sections (0.7 μm) were cut with a glass knife on an ultramicrotome and stained with DAPI and $DiOC_6$ to distinguish mitochondria ($DiOC_6$-positive) from plastids ($DiOC_6$-negative). Computer-aided reconstruction of serial images was performed using Cosmozone 2SA (Nikon Inc). Blue objects are nuclei; red objects are pt-nucleoids; white dots are mt-nucleoids; gray ribbons represent cell boundaries (partially visualized). Svn, sperm cell associated with vegetative nucleus; Sua, Sperm cell unassociated with vegetative nucleus; V, vegetative nucleus; S, sperm nuclei. Scale bar: 5 μm.

Contributors

Chieko Saito*, Department of Biological Sciences, Graduate School of Science, The University of Tokyo, 7-3-1 Hongo, Bunkyo-ku, Tokyo 113-0033, Japan
*E-mail: chiezo@bs.s.u-tokyo.ac.jp

References

13. Russel SD (1980) Participation of male cytoplasm during gamete fusion in an angiosperm, *Plumbago zeylanica*. Science 210:200–201. doi:10.1126/science.210.4466.200
14. Russel SD (1985) Preferential fertilization in *Plumbago zeylanica*: ultra-structural evidence for gamete-level recognition in an angiosperm. Proc Natl Acad Sci U S A 82:6129–6132
15. Saito C, Nagata N, Sakai A, Kuroiwa H, Kuroiwa T (2001) Behavior of plastid nucleoids during male gametogenesis in *Plumbago auriculata*. Protoplasma 216:143–154. doi:10.1007/BF02673866

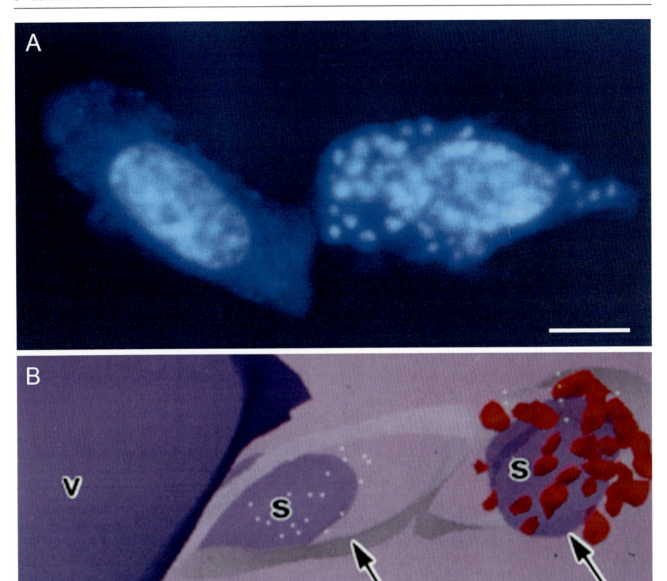

Plate 8.9

Pollen tube guidance toward the ovule

Male gametes of basal plants, such as sperms of sea algae and gymnosperms, are motile and able to swim toward female gametophytes. The sperm cells of angiosperms, on the other hand, lost motility during evolution and instead are carried to the female gametophyte by the pollen tube. The pollen tube is a single tubular cell that emerges from the pollen grain. After landing on the female stigma, the pollen grain hydrates and the pollen tube germinates from a pore on the surface of the pollen grain through the style by tip growth and reaches the ovary, where ovules are located. In the case of *Arabidopsis thaliana*, pollen tubes initially grow inside the transmitting tract, which is located at the center of the ovary (**A**).

When the pollen tube approaches the ovule, it changes its growth orientation and climbs up the funiculus, the tissue that connects the ovule and the transmitting tract. Guided by secreted signals from the female gametophyte, the pollen tube enters the female gametophyte through the micropyle of the ovule and releases sperm cells (arrowhead in **B**). Defensin-like cysteine-rich polypeptides, LUREs, have been identified as guidance molecules secreted from the synergid cells located adjacent to the egg cell in the female gametophyte in *Torenia fournieri* [16]. A mutation in *MYB98*, which encodes an R2R3-type MYB transcription factor, causes defects in secretion of this guidance signal and results in misguidance of the pollen tube in *Arabidopsis thaliana* [17]. In *myb98* mutants, the pollen tube often passes the micropyle of the mutant ovule and continues growing outside the ovule (arrowhead in **C**).

Self-pollinated *Arabidopsis thaliana* pistils were fixed with ethanol-acetic acid solution (ethanol: acetic acid = 9:1) overnight at room temperature, rinsed with an ethanol series (90 % ethanol for 20 m, 70 % for 20 m and 50 % for 5 m), and then treated with 1 N NaOH overnight at room temperature. Samples were subsequently stained with aniline blue solution (0.1 % aniline blue in 0.1 M K_3PO_4, pH 12.4) for more than 1 h to visualize pollen tubes. Whole pistil was mounted with 50 % glycerol and observed under UV fluorescent microscopy (ZEISS AxioImager.A2). Scale bars: 500 µm (**A**), 100 µm (**B, C**).

Contributors

Masahiro M. Kanaoka*, Division of Biological Science, Graduate School of Science, Nagoya University, Furo-cho, Chikusa-ku, Nagoya, Aichi 464-8602, Japan
*E-mail: mkanaoka@bio.nagoya-u.ac.jp

References

16. Okuda S, Tsutsui H, Shiina K, Sprunck S, Takeuchi H, Yui R, Kasahara RD, Hamamura Y, Mizukami A, Susaki D, Kawano N, Sakakibara T, Namiki S, Itoh K, Otsuka K, Matsuzaki M, Nozaki H, Kuroiwa T, Nakano A, Kanaoka MM, Dresselhaus T, Sasaki N, Higashiyama T (2009) Defensin-like polypeptide LUREs are pollen tube attractants secreted from synergid cells. Nature 458:357–361. doi:10.1038/nature07882
17. Kasahara RD, Portereiko MF, Sandaklie-Nikolova L, Rabiger DS, Drews GN (2005) MYB98 is required for pollen tube guidance and synergid cell differentiation in Arabidopsis. Plant Cell 17:2981–2992. doi:10.1105/tpc.105.034603

8 Generative Cells

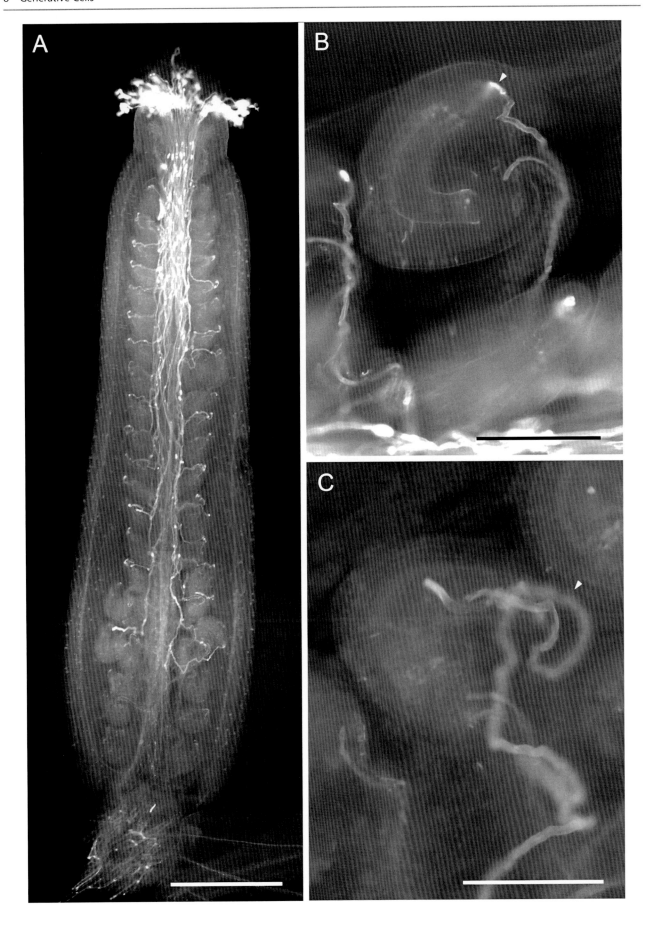

Plate 8.10

Semi-in vitro *Torenia* system for live-cell analysis of plant fertilization

Fertilization processes of flowering plants are hidden within the tissue layers of the pistil. A semi-in vitro system (or semi-in vivo system) was developed to study fertilization processes directly in the living material of *Torenia fournieri* [18, 19]. In this system, pollen tubes growing through a cut style are cultivated with excised ovules. Competent pollen tubes are attracted to the ovules, and pollen tube discharge and double fertilization occur. *T. fournieri* was used because it has a protruding embryo sac (female gametophyte). The egg cell, two synergid cells, and about half of the central cell are not covered with ovular integumentary cells. These gametophytic cells can be directly observed and manipulated in living material.

Establishment of this semi-in vitro system allows live study of plant fertilization and has led to two significant discoveries [20–22]. First, two synergid cells were identified as the source of the pollen tube attractant. Attractant molecules were finally identified as defensin-like peptides named LUREs after remaining elusive for over 140 years. The control of competency of the pollen tube by the female pistil was also discovered in the semi-in vitro system. Second, live-cell imaging of fertilization processes has provided many insights into the actual behavior of gametophytic cells. The semi-in vitro technique was applied in *Arabidopsis thaliana*, in which live-cell studies are now commonly performed. The behavior of two sperm cells and their nuclei during double fertilization were first captured in this semi-in vitro *Arabidopsis* system.

This figure is a dark-field image of the semi-in vitro *Torenia* system [19]. The green parts in each ovule are integument cells with chloroplasts. The diameter of the pollen tube is approximately 10 μm. Scale bar: 200 μm.

Contributors

Tetsuya Higashiyama*, Institute of Transformative Bio-Molecules (WPI-ITbM)/Division of Biological Science, Graduate School of Science/JST ERATO Higashiyama Live-Holonics Project, Nagoya University, Furo-cho, Chikusa-ku, Nagoya, Aichi 464-8602, Japan
*E-mail: higashi@bio.nagoya-u.ac.jp

References

18. Higashiyama T, Kuroiwa H, Kawano S, Kuroiwa T (1998) Guidance in vitro of the pollen tube to the naked embryo sac of *Torenia fournieri*. Plant Cell 10:2019–2031. doi:10.1105/tpc.10.12.2019

19. Higashiyama T, Yabe S, Sasaki N, Nishimura Y, Miyagishima S, Kuroiwa H, Kuroiwa T (2001) Pollen tube attraction by the synergid cell. Science 293:1480–1483. doi:10.1126/science.1062429

20. Higashiyama T, Hamamura Y (2008) Gametophytic pollen tube guidance. Sex Plant Reprod 21:17–26. doi:10.1007/s00497-007-0064-6

21. Takeuchi H, Higashiyama T (2011) Attraction of tip-growing pollen tubes by the female gametophyte. Curr Opin Plant Biol 14:614–621. doi:10.1016/j.pbi.2011.07.010

22. Hamamura Y, Nagahara S, Higashiyama T (2012) Double fertilization on the move. Curr Opin Plant Biol 15:70–77. doi:10.1016/j.pbi.2011.11.001

8 Generative Cells

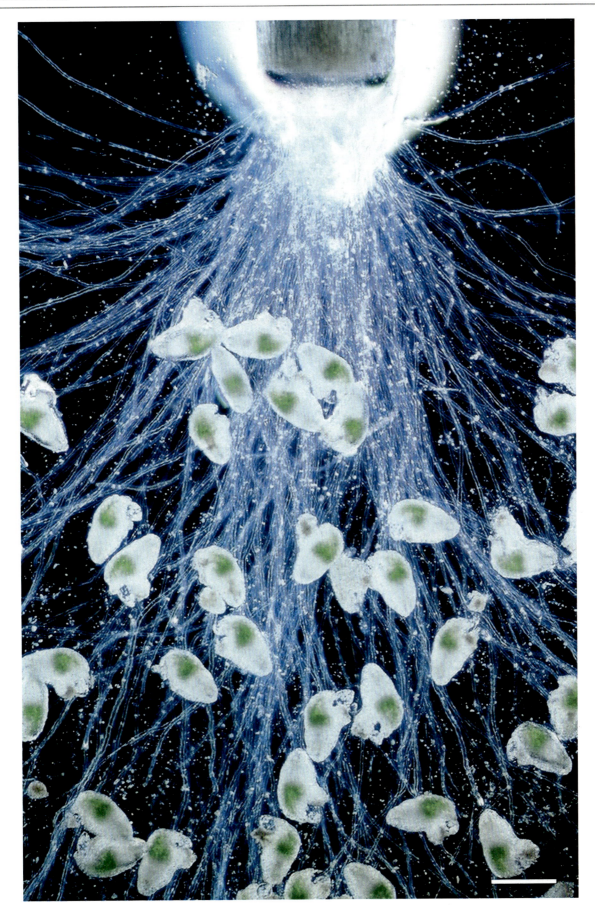

Plate 8.11

Protoplasts from plant female gametophytes

Typical Polygonum type plant female gametophytes consist of seven cells: one central cell, one egg cell, two synergid cells and three antipodal cells. The central cell is a diploid cell that is fertilized by a sperm cell to produce the endosperm. The egg cell is a haploid cell that is fertilized to produce the embryo. The two haploid synergid cells release small molecules to attract compatible pollen tubes into the ovule. Because of these important roles in plant sexual reproduction, the functions of these three types of cells have been studied cytologically and genetically. However, it is difficult to obtain pure samples of single cell types, because these gametophytic cells are deeply embedded in thick layers of ovule integuments.

One technique to isolate each cell type is laser-microdissection, which requires that samples must be fixed before dissection, and sectioning of the ovule is relatively difficult. Another technique is isolation of protoplasts. Plant species in the genus *Torenia* and *Lindernia* are especially good materials for this purpose because part of the female gametophyte protrudes from the ovule [23, 24]. A *Lindernia micrantha* ovule with its protruding embryo sac is shown (**inset**, arrowhead indicates embryo sac). After excision, cell wall digestion, and isolation of protoplasts, three types of gametophytic cells are easily distinguished by their morphological features (**main**). The diploid central cell protoplast is the biggest among them. Many cytoplasmic strands are seen in the egg cell (arrowhead). Synergid cells possess one large vacuole each (arrows).

Ovules were excised from pistil and treated with 300 μL of cellulase enzyme solution (1 % [w/v] Cellulase, 0.3 % Macerozyme RS [Yakult], 0.05 % Pectolyase Y-23 [Yakult], 5 mM $CaNO_3$ and 0.4 M Mannitol) for an hour at 28 °C to digest cell walls and isolate protoplasts. After cellulase treatment, gametophytic protoplasts were collected by PicoPipette (ALTAIR, Japan) under an inverted microscope (IX71, Olympus) for further analysis. Scale bars: 20 μm (**main**), 50 μm (**inset**).

Contributors

Masahiro M Kanaoka*, Division of Biological Science, Graduate School of Science, Nagoya University, Furo-cho, Chikusa-ku, Nagoya, Aichi 464-8602, Japan
*E-mail: mkanaoka@bio.nagoya-u.ac.jp

References

23. Kanaoka MM (2011) Technical note: isolation of intact gametophytic protoplasts from *Torenia* and *Lindernia* species. Cytologia 76:109. doi:10.1508/cytologia.76.109

24. Kawano N, Susaki D, Sasaki N, Higashiyama T, Kanaoka MM (2011) Isolation of gametophytic cells and identification of their cell-specific markers in *Torenia fournieri*, *T. concolor* and *Lindernia micrantha*. Cytologia 76:177–184. doi:10.1508/cytologia.76.177

8 Generative Cells

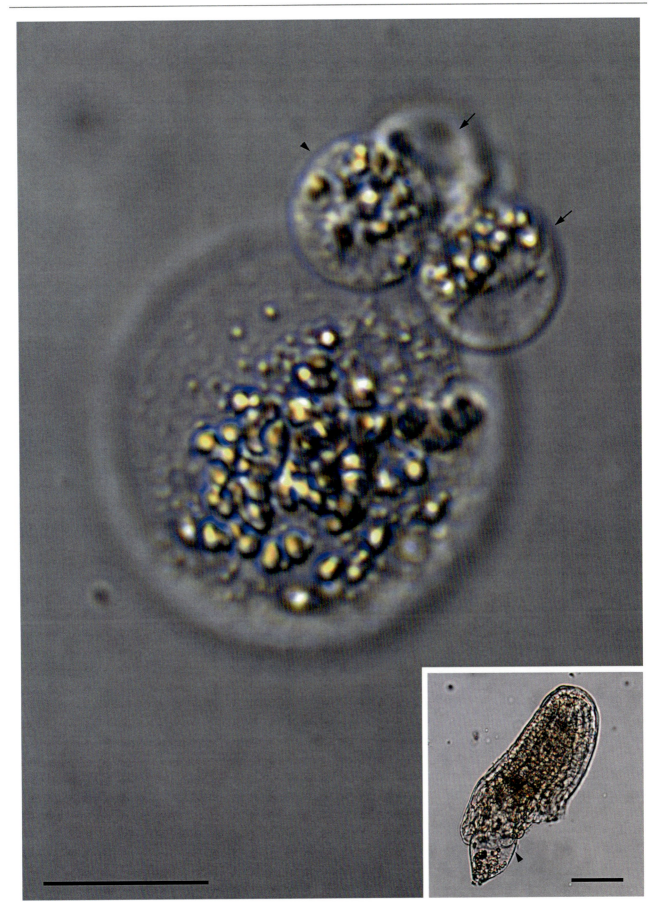

Plate 8.12

Egg cells with giant mitochondria in a higher plant, *Pelargonium zonale* Ait

Mature egg cells of *Pelargonium zonale* Ait. have giant mitochondria with large mitochondrial nuclei (mt-nuclei or nucleoids) [25]. The preferential development of giant mitochondria and mt-nuclei during the process of megasporogenesis and megagametogenesis was imaged by fluorescence microscopy at various developmental stages [26]. Reproductive cells at megaspore mother cell, meiosis, tetrad, and functioning megaspore stages contain many small mitochondria, each of which contains a small (about 0.3 μm in diameter), characteristic, uniformly stained mt-nucleus consisting of a small amount of DNA (0.3 Mbp). During the formation of the 2-, 4-, and 8-nucleate embryo sac, mt-nuclei do not markedly change their shapes or DNA contents. When cytokinesis has been completed within the embryo sac and differentiation of respective gametophytic cells has begun, mitochondria and mt-nuclei in the egg cell take on a small ring or string-like conformation. As embryo sac matures, mitochondria progressively enlarge and gradually become long, thick strings or stacks of concentric or half concentric rings. The embryo sac completely matures by flower opening.

In the mature embryo sac, a large balloon-shaped egg is situated with two synergid cells at the micropylar area. Mitochondrial nuclei within the mature egg cell are arranged as stacks of five to ten rings 5 to 7 μm in diameter (**A**). A close-up of one representative mt-nuclear complex with 9 stacks shows that there is no continuity between stacks as they are uniformly separated by about 0.3 μm (**B**). A 3D reconstruction of one mt-nuclear complex shows that nine mt-nuclei are assembled to compose one complex; while the innermost mt-nucleus is oval in shape, outer mt-nuclei are cup-shaped (**C**). Two mitochondrial-complexes, each of which is composed of double rings visualized by electron microscopy are also shown (**D**). Serial sectioning reveals that 44 mitochondrial complexes are present within an egg cell. Fluorometry using VIMPICS reveals that each stacked mitochondrial complex within a mature egg cell contains as much as 340–1,700 Mbp DNA, 40 times as much as one mitochondrion within the megaspore mother cell. Thus, a single egg cell contains at least 15,000 Mbp of mt-DNA, which is more DNA than the nucleus contains.

Samples were prepared by Technovit embedding, sectioning, and 4′,6-diamidino-2-phenylindole (DAPI) staining of ovules. Scale bars: 2.5 μm (**A**), 0.5 μm (**B–D**).

Contributors

Haruko Kuroiwa*, Tsuneyoshi Kuroiwa, CREST, Initiative Research Unit, College of Science, Rikkyo University, Toshima, Tokyo 171-8501, Japan
*E-mail: haruko-k@tvr.rikkyo.ne.jp

References

25. Kuroiwa H, Kuroiwa T (1992) Giant mitochondria in the mature egg cell of *Pelargonium zonale*. Protoplasma 168:184–188. doi:10.1007/BF01666264
26. Kuroiwa H, Ohta T, Kuroiwa T (1996) Studies on the development and three-dimensional reconstruction of giant mitochondria and their nuclei in egg cells of *Pelargonium zonale* Ait. Protoplasma 192:235–244. doi:10.1007/BF01273895

8 Generative Cells

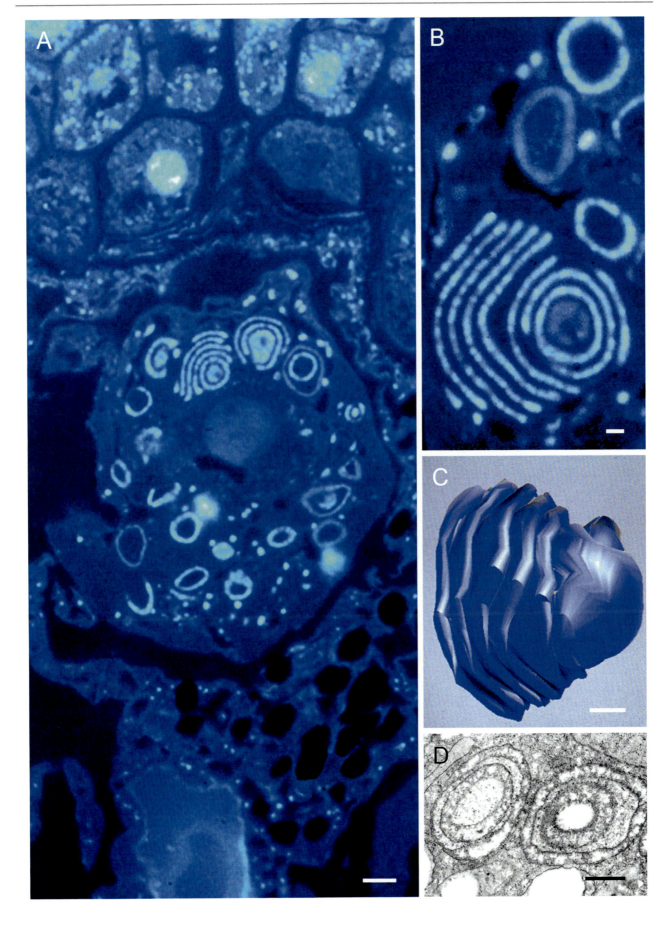

Plate 8.13

Zygote and sperm cells during early embryogenesis in a higher plant, *Pelargonium zonale* Ait

The angiosperm embryo sac is situated deep within several layers of integuments and integumentary tapetum cells. At the micropylar area of the embryo sac exists one egg cell, together with two triangular-shaped synergids [27]. The time from pollination to the first zygotic division is about 20–24 h. After pollination, one of the two synergids begins to degenerate prior to the arrival of the pollen tube to the embryo sac. By about 6 h after pollination, the pollen tube (PT) enters the degenerating, rather than the intact, synergid (**A**). At 6–9 h, two sperm cells migrate toward the egg cell and the central cell and respective cell fusions occur. By 9–15 h, nuclear fusion proceeds and the zygote nucleus moves to a central position in the cell. From 15 to 24 h, the zygote undergoes its first cell division and a two-celled proembryo is formed.

In the electron micrograph of the embryo sac 15 h after pollination, the centrally situated, balloon-shaped cell is a zygote which contains a large oval-shaped nucleus (N^{-Z}) with largely dispersed chromatin (**B**). Many vacuoles (V) appear around the nucleus and the zygote cell as a whole becomes vacuolated. Abundant lipid bodies appear adjacent to the zygote nucleus as an aggregation of dark, uniform-sized spherical bodies (~0.5 μm in diameter). Mitochondria have become a single ring (about 2 μm in diameter) or remain as assemblages of several rings which appear deformed or compressed among the vacuoles and dispersed randomly in the cytoplasm. Amyloplasts (Am) with large starch grains are also scattered throughout the cell. The triangular cell to the upper left of the zygote cell is the persistent synergid. In general, after entrance into the degenerated synergid and discharge of the first pollen tube, the invasion of other pollen tubes is interrupted and the zygote is protected from polyspermy. In this embryo sac, however, the persistent synergid appears to have received a second pollen tube because it contains a vegetative nucleus and two sperm cells. The vegetative nucleus (N) is irregular in shape and has dispersed chromatin. The sperm cell is surrounded by a unit membrane and has a spherical nucleus (N^{-S}). Small globular mitochondria (400 nm in diameter) and electron dense, oval-shaped plastids are scattered in the cytoplasm of the sperm cell. The degenerated synergid of the same embryo sac is observed in the serial section (**C**). The whole cell becomes extremely dark and it is impossible to distinguish organelles. At the chalazal region, a large hole, which is likely the path of the sperm cells, is observed (an asterisk).

The process of double fertilization in *P. zonale*, up to the formation of 2-celled proembryo was followed by Technovit-DAPI fluorescent microscopy and electron microscopy [28] to generate these images. Scale bars: 5 μm (**A**), 15 μm (**B, C**).

Contributors

Haruko Kuroiwa*, Tsuneyoshi Kuroiwa, CREST, Initiative Research Unit, College of Science, Rikkyo University, Toshima, Tokyo 171-8501, Japan
*E-mail: haruko-k@tvr.rikkyo.ne.jp

References

27. Kuroiwa H (1989) Ultrastructural examination of embryogenesis in *Crepis capillaris* (L.) Wallr: 1. The synergid before and after pollination. Bot Mag (Tokyo) 102:9–24. doi:10.1007/BF02488109
28. Kuroiwa H, Nishimura Y, Higashiyama T, Kuroiwa T (2002) *Pelargonium* embryogenesis: cytological investigations of organelles in early embryogenesis from the egg to the two-celled embryo. Sex Plant Reprod 15:1–12. doi:10.1007/s00497-002-0139-3

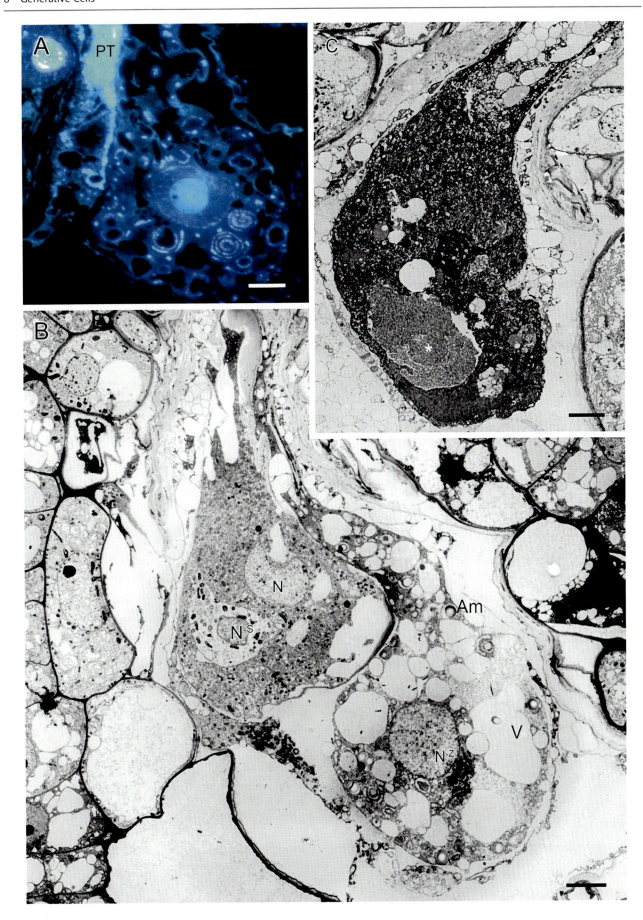

Plate 8.14

Cell geometry in a whole *Arabidopsis* seed visualized by X-ray micro-CT

Because rigid cell walls prevent cell movement during development, the origin of a plant cell determines its mature location in a tissue. Thus, precise determination of three-dimensional (3D) cell shapes and geometries would significantly contribute to understanding the correlation between cell shaping and global body shaping. X-ray computed tomography (CT) is one tool to examine the complex 3D organization of cells within tissues. X-ray micro-CT methods using X-rays from large synchrotron radiation facilities have achieved resolutions at the sub-micron level which enables examination of cellular architectures in tissues and organs. As the seed size of *Arabidopsis* is small enough (ca. 200 μm wide and ca. 500 μm long) to obtain whole seed images by X-ray micro-CT, one can clearly trace the outlines of almost all types of cells in a whole seed and can extract the facets and edges of each cell [29]. Using this method, cell geometrical analysis of a whole embryo is in progress.

A tomographic slice of a dry *Arabidopsis thaliana* (L.) Heynh. ecotype Columbia seed is shown here. In a longitudinal section view of a seed, outlines of epidermis, cortical cells and endodermal cells in hypocotyl-root axis and outlines of epidermis and parenchyma cells in cotyledon are clearly distinguishable (**A**). Although air spaces are frequently detected among cortical cells, they are never detected inside the procambium. A dissecting microscope image of a dry *Arabidopsis* seed is shown for comparison (**B**).

X-ray micro-CT was used to obtain a series of images. A seed was mounted on the top of a glass needle with cyanoacrylate glue, and the needle was placed on a rotation stage. The specimen-to-detector distance was maintained at 5 mm and the X-ray energy was adjusted to 8 keV. Images were recorded using a cooled CCD camera coupled with an optical lens and phosphor screen. The exposure time for each projection was 300 ms and a series of 900 projections were recorded over 180°. Effective pixel size was 0.5 μm. X-ray micro-CT was performed at the SPring-8 synchrotron radiation facility, beamline BL20XU, hutch No. 2 (Hyogo, Japan). Cot, cotyledon; Hy, hypocotyl; Ra, radicle. Scale bar: 50 μm.

Contributors

Yoshinobu Mineyuki[1]*, Aki Fukuda[2], Daisuke Yamauchi[1], Ichirou Karahara[3], [1]Department of Life Science, Graduate School of Life Science University of Hyogo, 2167 Shosha, Himeji, Hyogo 671-2280, Japan, [2]Department of Life Science, Faculty of Science, University of Hyogo, 2167 Shosha, Himeji, Hyogo 671-2280, Japan, [3]Department of Biology, Graduate School of Science and Engineering, University of Toyama, 3190 Gofuku, Toyama 930-8555, Japan
*E-mail: mineyuki@sci.u-hyogo.ac.jp

References

29. Yamauchi D, Tamaoki D, Hayami M, Uesugi K, Takeuchi A, Suzuki Y, Karahara I, Mineyuki Y (2012) Extracting tissue and cell outlines of Arabidopsis seeds using refraction contrast X-Ray CT at the SPring-8 facility. AIP Conf Proc 1466:237–242. doi:10.1063/1.4742298

8 Generative Cells

Chapter References

1. Kagami Y, Mogi Y, Arai T, Yamamoto M, Kuwano K, Kawano S (2008) Sexuality and uniparental inheritance of chloroplast DNA in the isogamous green alga *Ulva compressa* (Ulvophyceae). J Phycol 44:691–702. doi:10.1111/j.1529-8817.2008.00527.x
2. Mogi Y, Kagami Y, Kuwano K, Miyamura S, Nagumo T, Kawano S (2008) Asymmetry of eyespot and mating structure positions in *Ulva compressa* (Ulvales, Chlorophyta) revealed by a new FE-SEM method. J Phycol 44:1290–1299. doi:10.1111/j.1529-8817.2008.00573.x
3. Nishida H, Pigg KB, Rigby JF (2003) Swimming sperm in an extinct Gondwanan plant. Nature 422:396–397. doi:10.1038/422396a
4. Nishida H, Pigg KB, Kudo K, Rigby JF (2004) Zooidogamy in the late Permian genus *Glossopteris*. J Plant Res 117:323–328. doi:10.1007/s10265-004-0164-4
5. Nishida H, Pigg KB, Kudo K, Rigby JF (2007) New evidence of reproductive organs of *Glossopteris* based on permineralized fossils from Queensland, Australia. I. Ovulate organ *Homevaleia* gen nov. J Plant Res 120:539–549. doi:10.1007/s10265-007-0093-0
6. Hori T, Miyamura S (1997) Contribution to the knowledge of fertilization of gymnosperms with flagellated sperm cells: *Ginkgo biloba* and *Cycas revoluta*. In: Hori T, Ridge RW, Tulecke W, Tredici DP, Tremouillaux-Guiller J, Tobe H (eds) *Ginkgo biloba* a global treasure: from biology to medicine. Springer, Tokyo, pp 67–84. doi:10.1007/978-4-431-68416-9_6
7. Tanaka I (2011) Molecular morphological studies on pollen development using protoplasts. Plant Morphol 23:53–59
8. Tanaka I (1993) Development of male gametes in flowering plants. J Plant Res 106:55–63. doi:10.1007/BF02344373
9. Ueda K, Tanaka I (1995) Male gametic nucleus-specific H2B and H3 histones, designated gH2B and gH3, in *Lilium longiflorum*. Planta 197:289–295. doi:10.1007/BF00202649
10. Fujie M, Kuroiwa H, Kawano S, Mutoh S, Kuroiwa T (1994) Behavior of organelles and their nucleoids in the shoot apical meristem during leaf development in *Arabidopsis thaliana* L. Planta 194:395–405. doi:10.1007/BF00197541
11. Nagata N, Saito C, Sakai A, Kuroiwa H, Kuroiwa T (1999) The selective increase or decrease of organellar DNA in generative cells just after pollen mitosis one controls cytoplasmic inheritance. Planta 209:53–65. doi:10.1007/s004250050606
12. Fauré S, Noyer JL, Carreel F, Horry JP, Bakry F, Lanaud C (1994) Maternal inheritance of chloroplast genome and paternal inheritance of mitochondrial genome in bananas (*Musa acuminata*). Curr Genet 25:265–269. doi:10.1007/BF00357172
13. Russel SD (1980) Participation of male cytoplasm during gamete fusion in an angiosperm, *Plumbago zeylanica*. Science 210:200–201. doi:10.1126/science.210.4466.200
14. Russel SD (1985) Preferential fertilization in *Plumbago zeylanica*: ultra-structural evidence for gamete-level recognition in an angiosperm. Proc Natl Acad Sci U S A 82:6129–6132
15. Saito C, Nagata N, Sakai A, Kuroiwa H, Kuroiwa T (2001) Behavior of plastid nucleoids during male gametogenesis in *Plumbago auriculata*. Protoplasma 216:143–154. doi:10.1007/BF02673866
16. Okuda S, Tsutsui H, Shiina K, Sprunck S, Takeuchi H, Yui R, Kasahara RD, Hamamura Y, Mizukami A, Susaki D, Kawano N, Sakakibara T, Namiki S, Itoh K, Otsuka K, Matsuzaki M, Nozaki H, Kuroiwa T, Nakano A, Kanaoka MM, Dresselhaus T, Sasaki N, Higashiyama T (2009) Defensin-like polypeptide LUREs are pollen tube attractants secreted from synergid cells. Nature 458:357–361. doi:10.1038/nature07882
17. Kasahara RD, Portereiko MF, Sandaklie-Nikolova L, Rabiger DS, Drews GN (2005) MYB98 is required for pollen tube guidance and synergid cell differentiation in Arabidopsis. Plant Cell 17:2981–2992. doi:10.1105/tpc.105.034603
18. Higashiyama T, Kuroiwa H, Kawano S, Kuroiwa T (1998) Guidance in vitro of the pollen tube to the naked embryo sac of *Torenia fournieri*. Plant Cell 10:2019–2031. doi:10.1105/tpc.10.12.2019
19. Higashiyama T, Yabe S, Sasaki N, Nishimura Y, Miyagishima S, Kuroiwa H, Kuroiwa T (2001) Pollen tube attraction by the synergid cell. Science 293:1480–1483. doi:10.1126/science.1062429
20. Higashiyama T, Hamamura Y (2008) Gametophytic pollen tube guidance. Sex Plant Reprod 21:17–26. doi:10.1007/s00497-007-0064-6
21. Takeuchi H, Higashiyama T (2011) Attraction of tip-growing pollen tubes by the female gametophyte. Curr Opin Plant Biol 14:614–621. doi:10.1016/j.pbi.2011.07.010
22. Hamamura Y, Nagahara S, Higashiyama T (2012) Double fertilization on the move. Curr Opin Plant Biol 15:70–77. doi:10.1016/j.pbi.2011.11.001
23. Kanaoka MM (2011) Technical note: isolation of intact gametophytic protoplasts from *Torenia* and *Lindernia* species. Cytologia 76:109. doi:10.1508/cytologia.76.109
24. Kawano N, Susaki D, Sasaki N, Higashiyama T, Kanaoka MM (2011) Isolation of gametophytic cells and identification of their cell-specific markers in *Torenia fournieri*, *T. concolor* and *Lindernia micrantha*. Cytologia 76:177–184. doi:10.1508/cytologia.76.177
25. Kuroiwa H, Kuroiwa T (1992) Giant mitochondria in the mature egg cell of *Pelargonium zonale*. Protoplasma 168:184–188. doi:10.1007/BF01666264
26. Kuroiwa H, Ohta T, Kuroiwa T (1996) Studies on the development and three-dimensional reconstruction of giant mitochondria and their nuclei in egg cells of *Pelargonium zonale* Ait. Protoplasma 192:235–244. doi:10.1007/BF01273895
27. Kuroiwa H (1989) Ultrastructural examination of embryogenesis in *Crepis capillaris* (L.) Wallr: 1. The synergid before and after pollination. Bot Mag (Tokyo) 102:9–24. doi:10.1007/BF02488109
28. Kuroiwa H, Nishimura Y, Higashiyama T, Kuroiwa T (2002) *Pelargonium* embryogenesis: cytological investigations of organelles in early embryogenesis from the egg to the two-celled embryo. Sex Plant Reprod 15:1–12. doi:10.1007/s00497-002-0139-3
29. Yamauchi D, Tamaoki D, Hayami M, Uesugi K, Takeuchi A, Suzuki Y, Karahara I, Mineyuki Y (2012) Extracting tissue and cell outlines of Arabidopsis seeds using refraction contrast X-Ray CT at the SPring-8 facility. AIP Conf Proc 1466:237–242. doi:10.1063/1.4742298

Meristems

9

Hirokazu Tsukaya

The meristem is a somewhat enigmatic concept. Esau (1960) described them as "embryonic tissue zones, the *meristems*, in which the addition of new cells continues." The term "embryonic" is not clear or strict, but shoot and root apical meristems have been thought to be the meristems defined here. Most molecular developmental biologists narrowly adopt a definition of meristems as proliferating tissues that maintains self-renewing stem cells under control of the WUSCHEL-RELATED HOMEOBOX (WOX) gene family. In fact, at least in the *Arabidopsis thaliana* (L.) Heynh., it is well known that *WUSCHEL* and *WOX5* are expressed in shoot and apical meristems, respectively, and regulate stem cell maintenance. In this sense, the lateral meristem (or cambium tissue) is also a meristem since *WOX4* is known to function in controlling the self-renewal of stem cells. On the other hand, "meristematic tissues" such as intercalary meristem in leaves or stems are excluded from this definition at present because WOX gene expression or stem cells have not been observed in these tissues. However, intercalary meristems are indeed actively meristematic, or cell-proliferating. Similarly, "meristemoid" is also a complex concept. In a meristemoid lineage, both self renewal and stem cell identity are recognized, but WOX is not known to have a role in these cells in *A. thaliana* at least at present. Is meristemoid a form of meritem? The above confusion reveals that the definition of stem cells in plant science is not fully mature. The tissues in which WOX members function have not yet been fully identified, and we cannot exclude intercalary meristems from such definition at present.

Conscious on the above complications, we overview a range of apical meristems from whence postembryonic cells and tissues are derived in this chapter. First, R. Imaitchi shows meristems of the fern gametophyte which is a freely living haploid phase. Recent studies have revealed that genetic regulation of organogenesis is very different in the haploid phase than in the diploid phase. The apical cell maintenance system is also known to have unique features and molecular regulatory systems are not yet known.

R. Imaichi also supplies various images of fern, lycophyte and angiopserm shoot and root apical meristems. Fern and lycophyte shoots and roots are derived from a single apical cell, while angiosperms develop highly organized multicellular tissues in their apical meristems, as described in detail here. The evolution of these structures is one of facsinating aspect of plant developmental biology and plant evolutionary biology.

The "leaf meristem," a type of intercalary meristem, is also described in this chapter. Leaf meristems are very active and supply huge amount of cells which are generated during a certain period of activity. Ichihashi et al. show the leaf meristem of *A. thaliana* by a modified mPS-PI method. A combination of an EdU-detection system to identify DNA-replicating cells with the modified mPS-PI might help us to re-examine the nature of the leaf meristem and understand whether WOX genes function in the leaf meristem and if the leaf meristem is indeed different from apical meristems in terms of stem cells.

Finally, Takagi et al. present asymmetric cell division visualized by GFP-marked gene expression pattern in the root apical meristem of *A. thaliana*. The presence of well-organized cell lineages in the root apical meristem has fascinated scientists, and many key genetic factors have been identified which regulate organogenesis. More new discoveries will likely arise from this model system.

H. Tsukaya (✉)
Department of Biological Sciences, Graduate School of Science,
The University of Tokyo, Bunkyo-ku, Tokyo 113-0033, Japan
e-mail: tsukaya@bs.s.u-tokyo.ac.jp

T. Noguchi et al. (eds.), *Atlas of Plant Cell Structure*,
DOI 10.1007/978-4-431-54941-3_9, © Springer Japan 2014

Plate 9.1

Two types of meristem involved in development of the fern gametophyte

The fern gametophyte is generally cordate in shape and develops from the activity of two different meristems, the apical-cell based and multicellular meristems. Both meristems are shown here in the same growing gametophyte of *Histiopteris incisa* (Thunb.) J. Sm. (Dennstaedtiaceae). **A**, **B**, and **C** are epi-iluminated micrographs taken 25, 32, and 47 days after start of culture, respectively. **D** is a line drawing of **B**.

Early in gametophyte development, the triangular apical cell forms and continues to cut off segments from the two lateral faces (arrow) (**A**). After producing several segments, the apical cell undergoes periclinal and anticlinal divisions (arrow in **B**; red cells in **D**), and disappears. Segments produced by the apical cell give rise to large cell packets (differently colored areas in **D**) by cell division, contributing to gametophyte growth. After disappearance of the apical cell, a row of elongate rectangular cells, i.e. the multicellular meristem, form in the notch of the gametophyte (arrowhead) (**C**). Asterisks in **C** indicate archegonia.

Spores were sterilized using 0.5 % sodium hypochloride and were sown on 1 % solidified agar medium with Parker and Thompson's micronutrients. Cultures were grown under continuous white fluorescent illumination at 24 °C. Microscopic images of growing gametophytes were captured using a metallurgical microscope with epi-illumination every 24 h [1]. Scale bars: 100 μm.

Contributors

Ryoko Imaichi*, Department of Chemical and Biological Sciences, Japan Women's University, 2-8-1 Mejirodai, Tokyo 112-8681, Japan
*E-mail: ryoko@fc.jwu.ac.jp

References

1. Takahashi N, Hashino M, Kami C, Imaichi R (2009) Developmental morphology of strap-shaped gametophytes of *Colysis decurrens*: a new look at meristem development and function in fern gametophytes. Ann Bot 104:1353–1361

9 Meristems

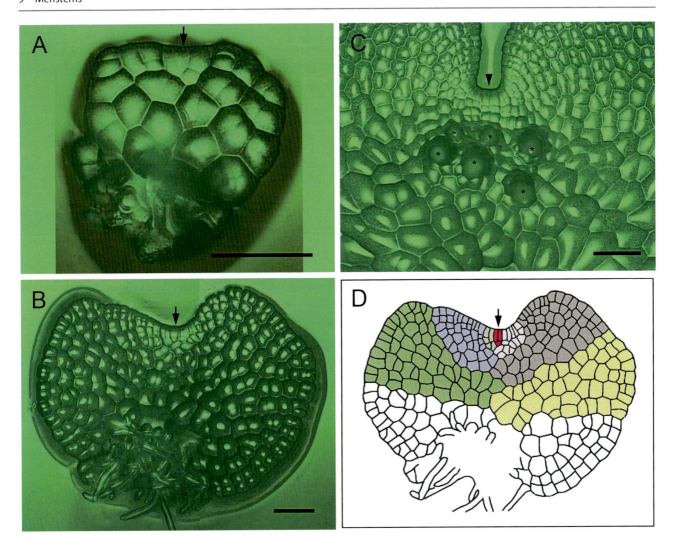

Plate 9.2

Structures of fern and lycophyte shoot apical meristems (SAMs)

Shoot apical meristem types vary across vascular plants taxa. Ferns and some lycophytes (Selaginellaceae) have monoplex SAMs, other lycophytes (Isoetaceae, Lycopodiaceae) and gymnosperms have simplex SAMs, and angiosperms have duplex SAMs. Monoplex SAMs are characterized by possessing a single apical cell, the simplex SAMs by plural initial cells in one zone, and duplex SAMs by plural initial cells in two zones.

In a median longitudinal section of a monoplex SAM of *Dicranopteris linearis* (Burm.f.) Underw. (Gleicheniaceae), a triangular apical cell is distinguished (arrow in **A**). The apical cell cuts off derivative cells (segments) from its three lateral facets. Each segment undergoes periclinal and anticlinal divisions to contribute all cells of a stem. In a median longitudinal section of a simplex SAM of *Huperzia serrata* (Thunb.) Trevis. (Lycopodiaceae), however, a triangular apical cell is not distinguished, and instead a group of small quadrangle cells occupies the tip of a SAM (**B**).

Small pieces of shoot tips including SAM and leaf primordia were cut off under a dissecting microscope and fixed with FAA, dehydrated through a graded ethanol series, and embedded in Technovit 7100 resin. Sections were longitudinally cut at a thickness of 2 μm and stained with 0.2 % Toluidine blue in 0.2 % sodium tetraborate, and 2 % Orange G-5 % Tannic Acid (a modified Sharman's stain). Light microscopic images were taken by OLYMPUS DP70. Scale bars: 50 μm.

Contributors

Ryoko Imaichi*, Department of Chemical and Biological Sciences, Japan Women's University, 2-8-1 Mejirodai, Tokyo 112-8681, Japan
*E-mail: ryoko@fc.jwu.ac.jp

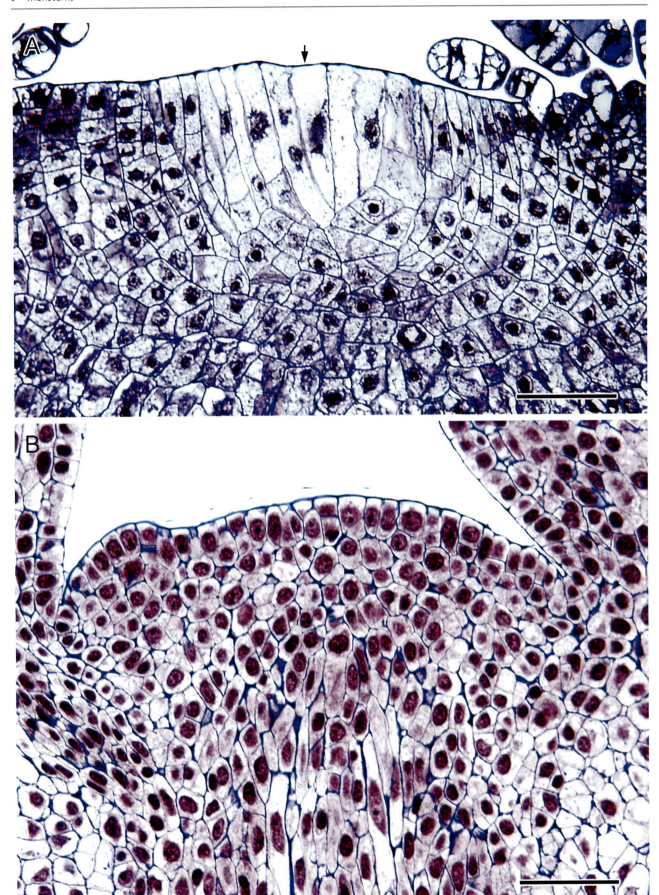

Plate 9.3

Structures of angiosperm SAMs

Angiosperm SAMs show a layered structure, the so-called tunica-corpus structure. Cells in the tunica layer undergo cell divisions only in the anticlinal orientation, whereas cells in the corpus undergo divisions in various orientations. The tunica and corpus have initial cells independently in two tiers, and are designated as duplex SAMs. Median longitudinal sections of *Ternstroemia gymnanthera* (Wight et Arn.) Bedd. (Theaceae) and *Aucuba japonica* Thunb. var. japonica (Cornaceae) SAMs are shown in **A** and **B**, respectively.

The dome-shaped SAM of *Ternstroemia gymnanthera* shows tunica-corpus construction with a two-layered tunica (**A**). The SAM of *Aucuba japonica* is very flat in outline, but it also shows tunica-corpus construction (**B**). Some zones are cytohistologically recognizeable here, which are comparable with gymnosperm SAM zones. A group of lightly stained, relatively large cells occupy the center of the SAM, akin to a central mother cell zone. Small, densely stained, cells occupy the apical flank, similar to the peripheral zone. Below the central mother cell zone, piles of flat rectangular cells are present, which resemble rib meristem.

Shoot tips of both species were cut off under a dissecting microscope, and fixed with FAA, dehydrated through a graded ethanol series, embedded in Technovit 7100 resin. Median longitudinal sections (2 μm thick) were cut and stained with a modified Sharman's stain. Light microscopic images were taken by OLYMPUS DP70. Scale bars: 50 μm.

Contributors

Ryoko Imaichi*, Department of Chemical and Biological Sciences, Japan Women's University, 2-8-1 Mejirodai, Tokyo 112-8681, Japan
*E-mail: ryoko@fc.jwu.ac.jp

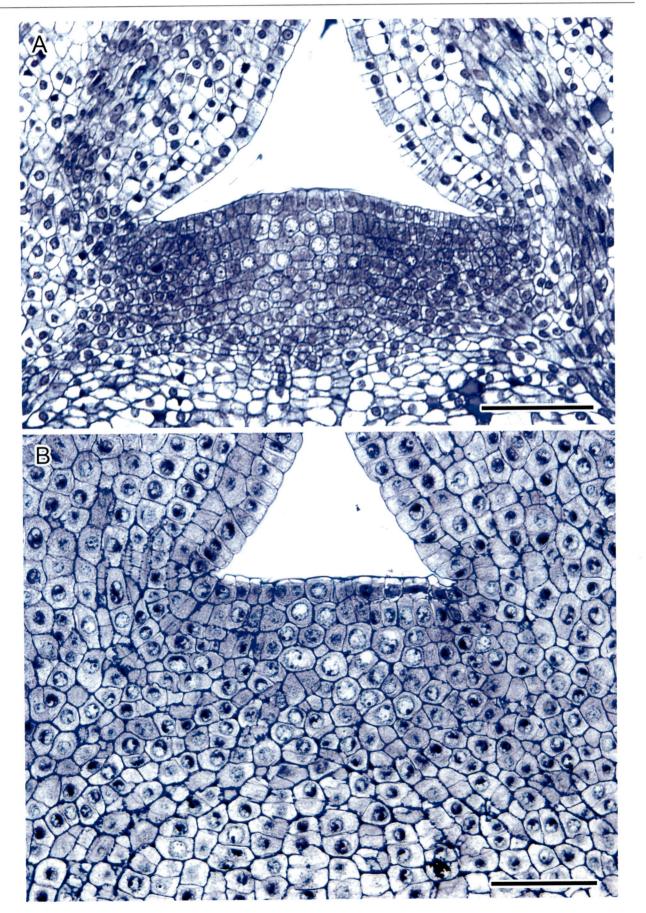

Plate 9.4

Arabidopsis thaliana leaf blade and leaf petiole

Leaves are the most fundamental units of organogenesis in seed plants. The basic structures of a leaf are the leaf blade and the leaf petiole. Studies investigating how structures differentiate from a leaf primordium are lacking. Ichihashi et al. [2] improved the modified pseudo-Schiff propidum iodide (mPS-PI) technique reported by Truernit et al. [3] and observed the development of leaf blade and leaf petiole of *Arabidopsis thaliana* (L.) Heynh. The leaf primordia has differences in cell shape and arrangement between the tip and base of the leaf primordium 3 days after sowing. This developmental stage has mitotic cells distributed almost uniformly in leaf primordia. This is the earliest developmental event identified related to the differentiation of the leaf blade and leaf petiole. The apical part of the leaf primordium has small, nonpolarized cells with longitudinal, transverse, and oblique arrangements of cross walls relative to the proximal-distal axis. On the other hand, the basal part of the leaf primordium has large, longitudinally polarized cells arranged parallel to the proximal-distal axis. After differentiation of the proximal-distal regions, cell proliferation in the leaf primordium appears to accelerate, and the morphology of the leaf blade/petiole junction becomes more conspicuous. Thus the regulation in growth direction might be an initial cue to leaf structure. The figures shown here are mPS-PI-stained shoot apex (**A**) and leaf primordia (**B**). A dual-imaging method to visualize cell walls and nuclei in S-phase cells in combination with the EdU-detection technique [4] can precisely depict the distribution of proliferating cells in leaf primordia (**C**).

Materials were fixed in 50 % methanol and 10 % acetic acid at 4 °C for more than 12 h. Samples were then treated with 80 %, 90 %, and 100 % (v/v) ethanol at 80 °C for 5 min each; then transferred to 50 % methanol and 10 % acetic acid and incubated for 1 h. Samples were rinsed with water and incubated in 1 % periodic acid at room temperature for 40 min; rinsed again with water and incubated in 1 mg/mL propidium iodide aqueous solution for 5 min. Confocal microscopic observations were carried out after transferring samples onto slides with a chloral hydrate solution for 30 min. Scale bars: 50 μm.

Contributors

Yasunori Ichihashi[1], Kensuke Kawade[2], Hirokazu Tsukaya[3]*, [1]Department of Plant Biology, University of California at Davis, One Shields Avenue, Davis, CA 95616, USA, [2]RIKEN CSRS, 1-7-22 Suehiro-cho, Tsurumi, Yokohama, Kanagawa 230-0045, Japan, [3]Department of Biological Sciences, Graduate School of Science, The University of Tokyo, Bunkyo-ku, Tokyo 113-0033, Japan
*E-mail: tsukaya@bs.s.u-tokyo.ac.jp

References

2. Ichihashi Y, Kawade K, Usami T, Horiguchi G, Takahashi T, Tsukaya H (2011) Key proliferative activity in the junction between the leaf blade and leaf petiole of Arabidopsis. Plant Physiol 157:1151–1162
3. Truernit E, Bauby H, Dubreucq B, Grandjean O, Runions J, Barthélémy J, Palauqui JC (2008) High-resolution whole-mount imaging of three-dimensional tissue organization and gene expression enables the study of phloem development and structure in Arabidopsis. Plant Cell 20:1494–1503
4. Kotogány E, Dudits D, Horváth GV, Ayaydin F (2010) A rapid and robust assay for detection of S-phase cell cycle progression in plant cells and tissues by using ethynyl deoxyuridine. Plant Methods 6:5

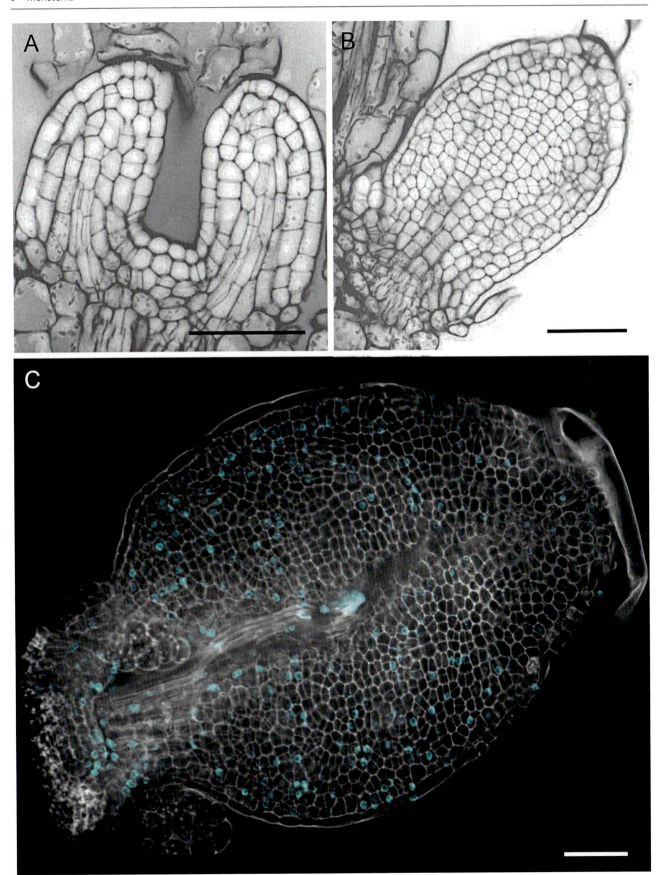

Plate 9.5

Structures of fern and lycophyte root apical meristems (RAMs)

Like SAMs, root apical meristem types also vary across vascular plants. However, RAMs show a wider variety of structures than SAMs. Median longitudinal sections of a fern, *Hypolepis punctata* (Thunb.) Mett. ex Kuhn (Dennstaedtiaceae), and a lycophyte, *Huperzia serrata* (Thunb.) Trevis. (Lycopodiaceae), are shown in A and B, respectively.

In *Hypolepis punctata* RAM, the tetrahedral apical cell is very apparent, and merophytes of different size are regularly arranged around the apical cell, indicating that all cells of the root, including root cap cells are derived from the apical cell (**A**). In contrast, *Huperzia serrata* RAM have a layered organization rather than a single apical cell (**B**). It appears that the epidermis, cortex, central cylinder, and root cap have independent initial cells.

Root tips of *Hypolepis punctata* were fixed with 2.5 % glutalaldehyde (GA) plus 1.0 % paraformaldehyde (PFA) in 0.1 M phosphate buffer and were postfixed with 2 % osmium tetroxide in 0.1 M phosphate buffer. Fixed samples were dehydrated in an ethanol–acetone series and embedded in epoxy resin (Quetol-653 resin, Nisshin EM), and cut into 2 μm thick longitudinally sections. Sections were stained with 0.2 % Toluidine blue and 0.5 % Sodium tetraborate decahydrate. Root tips of *Huperzia serrata* were fixed with FAA, and dehydrated through a graded ethanol series, and embedded in Technovit 7100 resin. 2 μm thick median longitudinal sections were cut and stained with a modified Sharman's stain. Light microscopic images were taken by OLYMPUS DP70. Scale bars: 50 μm.

Contributors

Ryoko Imaichi*, Department of Chemical and Biological Sciences, Japan Women's University, 2-8-1 Mejirodai, Tokyo 112-8681, Japan
*E-mail: ryoko@fc.jwu.ac.jp

9 Meristems 197

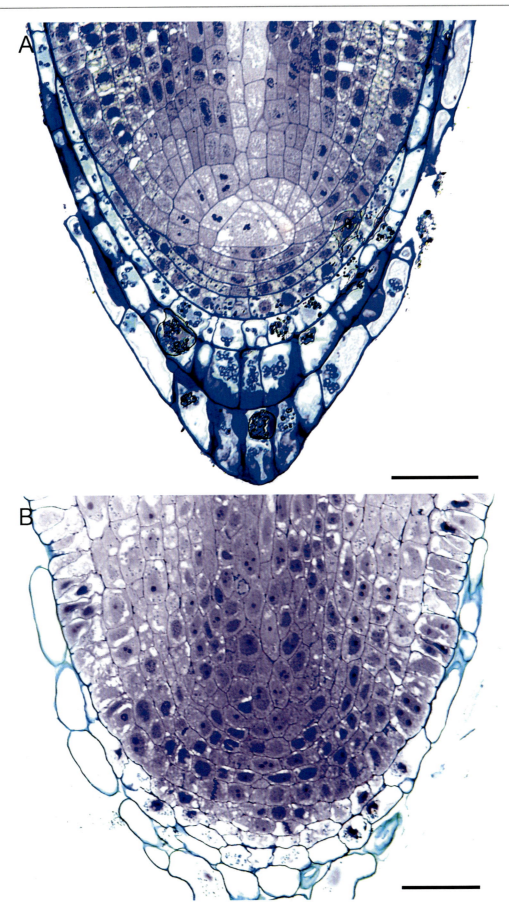

Plate 9.6

Structures of angiosperm RAMs

Angiosperm RAMs can be classified into two types, open and closed, although several subtypes have been described [5, 6]. In the open type, all cells of a root including the rootcap are derived from common initial cells. In the closed type, the central cylinder, cortex, and rootcap are traceable to independent layers of cells in the apical meristem. Two open type RAMs are shown here: the most basal taxon of primitive angiosperms, *Amborella trichopoda* Baill. (Amborellaceae) (**A**), and a eurosid taxon, *Pisum sativum* L. (Fabaceae) (**B**).

The *Amborella trichopoda* RAM has a small number of common initials, from which the central cylinder, cortex and epidermis, and rootcap are all derived (**A**). The *Pisum sativum* RAM is much larger than that of *Amborella trichopoda*, and is characterized by a plate of initial cells across the tips of the central cylinder, cortex and base of the columella (**B**). This type of RAM was specially designated the basic-open type by Groot et al. [5], or the open transverse eudicot (OTuD) type by Heisch and Seago [6].

Root tips were cut off under a binocular dissecting microscope, fixed in 2.5 % glutalaldehyde (GA) in 0.1 % phosphate buffer (for *Pisum sativum*), or 2.5 % glutalaldehyde (GA) plus 1.0 % paraformaldehyde (PFA) in phosphate buffer (for *Amborella trichopoda*), then postfixed with 2 % osmium tetroxide in 0.1 M phosphate buffer. Fixed samples were dehydrated in an ethanol–acetone series and embedded in epoxy resin (Quetol-653 resin, Nisshin EM, Tokyo), and cut into longitudinal sections 2 μm thick. Sections were stained with 0.2 % Toluidine blue and 0.5 % Sodium tetraborate decahydrate. Light microscopic images were taken by OLYMPUS DP70. Scale bars: 50 μm.

Contributors

Ryoko Imaichi*, Department of Chemical and Biological Sciences, Japan Women's University, 2-8-1 Mejirodai, Tokyo 112-8681, Japan
*E-mail: ryoko@fc.jwu.ac.jp

References

5. Groot EP, Doyle JA, Nichol SA, Rost TL (2004) Phylogenetic distribution and evolution of root apical meristem organization in dicotyledonous angiosperms. Int J Plant Sci 165:97–105
6. Heimsch C, Seago JL Jr (2008) Organization of the root apical meristsem in angiosperms. Am J Bot 95:1–21

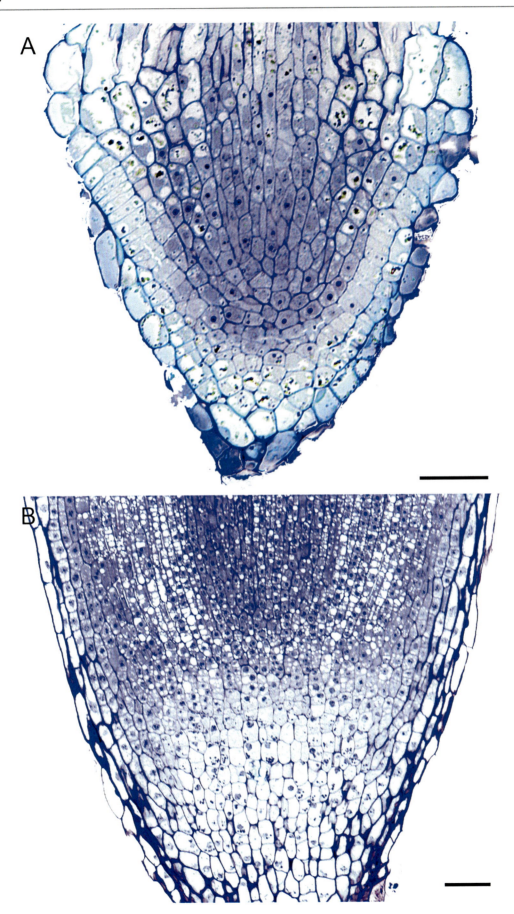

Plate 9.7

Asymmetric cell division forms endodermis and cortex in *Arabidopsis thaliana* root

Arabidopsis thaliana roots are typically reported to have four layers surrounding the central vascular tissue. The pericycle, endodermis, cortex, and epidermis are vertically arranged around the enteral vascular tissue. The cortex and endodermis are each one layer in *Arabidopsis thaliana*. Asymmetric cell division contributes to the establishment and propagation of the cellular pattern of plant tissues. The *SHORT-ROOT* (*SHR*) gene plays an important role in asymmetric cell division, in which the endodermis and cortex are differentiated and the endodermis in the root is specified [7]. SHR localizes to both nuclei and cytoplasm in the stele and to nuclei in the quiescent centre (QC), cortex/endodermis initial and daughter cells, and all cells of the endodermis. SHR transcripts can be found only in the stele, whereas the SHR gene seems not to be expressed in any of these cells in which it is localized only to the nucleus, as confirmed both by in situ hybridization and indirect detection with SHR promoter-reporter fusion gene. The SHR protein probably moves between cells and its translocation may produce non-cell-autonomous activity in root radial patterning [8].

The root cap in this image is stained with propidium iodide (red). Green fluorescence derived from SHR fused with green fluorescence protein (GFP) is detected in both nuclei and cytoplasm of the stele. In contrast, fluorescence is specifically detected in the nuclei of QC, the cortex/endodermis initial and daughter cells, and endodermis. Collumera cells of the root cap contain amyloplasts which have starch grains.

After sterilizing the seeds and imbibing them in the dark at 4 °C for 1 day. The plates were sealed with surgical tape and incubated them in a nearly vertical position in a growth chamber at 22 °C with a 16 h-light/8 h-dark photoperiod for 2 days. The coding region of SHR was fused with GFP under the SHR own promoter. GFP fluorescence (green) and propidium-iodide-stained cell walls (red) were observed under an inverted fluorescent microscope (IX-81, Olympus) equipped with a confocal scanning unit (CSUX-1, Yokogawa) and a sCMOS camera (Neo 5.5sCMOS, ANDOR). Images were analyzed with ImageJ software. Scale bar: 50 μm.

Contributors

Mai Takagi, Sachihiro Matsunaga*, Department of Applied Biological Science, Faculty of Science and Technology, Tokyo University of Science, 2641 Yamazaki, Noda, Chiba 278-8510, Japan
*E-mail: sachi@rs.tus.ac.jp

References

7. Helariutta Y, Fukaki H, Wysocka-Diller J, Nakajima K, Jung J, Sena G, Hauser MT, Benfey PN (2000) The *SHORT-ROOT* gene controls radial patterning of the *Arabidopsis* root through radial signaling. Cell 101:555–567. doi:10.1016/S0092-8674(00)80865-X
8. Nakajima K, Sena G, Nawy T, Benfey PN (2001) Intercellular movement of the putative transcription factor SHR in root patterning. Nature 413:307–311. doi:10.1038/35095061

9 Meristems

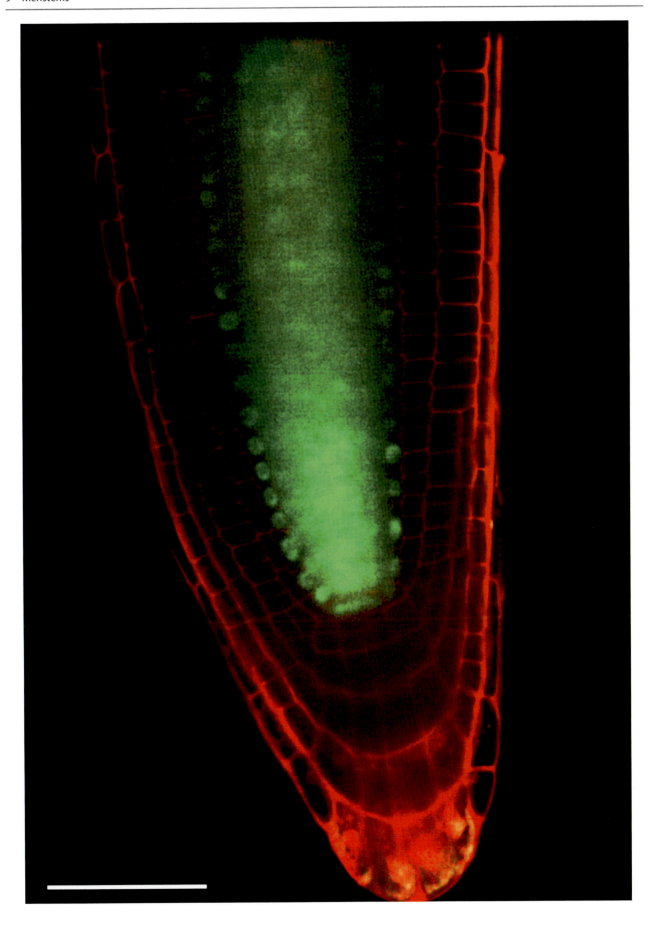

Chapter References

1. Takahashi N, Hashino M, Kami C, Imaichi R (2009) Developmental morphology of strap-shaped gametophytes of *Colysis decurrens*: a new look at meristem development and function in fern gametophytes. Ann Bot 104:1353–1361
2. Ichihashi Y, Kawade K, Usami T, Horiguchi G, Takahashi T, Tsukaya H (2011) Key proliferative activity in the junction between the leaf blade and leaf petiole of Arabidopsis. Plant Physiol 157:1151–1162
3. Truernit E, Bauby H, Dubreucq B, Grandjean O, Runions J, Barthélémy J, Palauqui JC (2008) High-resolution whole-mount imaging of three-dimensional tissue organization and gene expression enables the study of phloem development and structure in Arabidopsis. Plant Cell 20:1494–1503
4. Kotogány E, Dudits D, Horváth GV, Ayaydin F (2010) A rapid and robust assay for detection of S-phase cell cycle progression in plant cells and tissues by using ethynyl deoxyuridine. Plant Methods 6:5
5. Groot EP, Doyle JA, Nichol SA, Rost TL (2004) Phylogenetic distribution and evolution of root apical meristem organization in dicotyledonous angiosperms. Int J Plant Sci 165:97–105
6. Heimsch C, Seago JL Jr (2008) Organization of the root apical meristsem in angiosperms. Am J Bot 95:1–21
7. Helariutta Y, Fukaki H, Wysocka-Diller J, Nakajima K, Jung J, Sena G, Hauser MT, Benfey PN (2000) The *SHORT-ROOT* gene controls radial patterning of the *Arabidopsis* root through radial signaling. Cell 101:555–567. doi:10.1016/S0092-8674(00)80865-X
8. Nakajima K, Sena G, Nawy T, Benfey PN (2001) Intercellular movement of the putative transcription factor SHR in root patterning. Nature 413:307–311. doi:10.1038/35095061